THE
EFFICIENCY
TRAP

Finding a Better Way to Achieve a

SUSTAINABLE ENERGY FUTURE

STEVE HALLETT

 **Prometheus Books**

59 John Glenn Drive
Amherst, New York 14228–2119

Published 2013 by Prometheus Books

Cover image © 2013 Maciej Frolow/Media Bakery
Cover design by Jacqueline Nasso Cook

Inquiries should be addressed to
Prometheus Books
59 John Glenn Drive
Amherst, New York 14228–2119
VOICE: 716–691–0133
FAX: 716–691–0137
WWW.PROMETHEUSBOOKS.COM

17 16 15 14 13   5 4 3 2 1

Library of Congress Cataloging-in-Publication Data Pending

ISBN 978–1–61614–725–9 (pbk. : alk. paper)
ISBN 978–1–61614–726–6 (ebook)

Printed in the United States of America on acid-free paper

To Dan, Pat, and Sami

CONTENTS

THE EIGHTH DEADLY SIN

> I believe that one ought to have only as much market efficiency
> as one needs, because everything that we value in human life
> is within the realm of inefficiency—love, family, attachment,
> community, culture, old habits, comfortable old shoes.
>
> —Edward Luttwak, *Turbo-Capitalism:*
> *Winners and Losers in the Global Economy*[1]

There is very little ambiguity about the fact that efficiency is a virtue. The highest praise a letter of recommendation might contain is likely to be "So-and-So has been one of our best employees: she is very efficient." I feel good when I do something efficiently, such as finishing a report on time or leaving the house early enough in the morning so I don't have to drive like a bat out of hell—wasting gas—to get to work on time. I feel downright heroic if I remember to buy beer on the way home so that I don't have to go out again. I feel virtuous when I'm efficient enough to keep a clean towel in the bathroom so I don't have to waste time running naked and wet through the house looking for one. I feel doubly virtuous if I'm efficient enough to hang the thing up after I've used it (a rare occurrence, alas) so that it can dry.

Efficiency is prized so highly that inefficiency, its evil twin, might well qualify as the eighth deadly sin. A coworker is likely to be gossiped about at the water cooler for being inefficient. "Can you believe how long it took him to write that report?" Governments can be famously inefficient, and they're often despised for it. Perhaps the construction of a local road ran over budget and was finished a year late: unforgivable inefficiency. One of the excuses for dispossessing millions of people from their land by "pioneers" has been that the native residents were standing in the way of progress by not using the land efficiently.

The Latin word *efficere* is the root of the words *efficiency*, *effectiveness*, and *efficacy*, words that are often used interchangeably but do not have the same meanings at all. Effectiveness and efficacy simply refer to how well something works. Efficiency is the amount of work generated by a given amount of an input.

Engineers are not confused by the term *efficiency*. To an engineer, efficiency is the amount of work generated per unit of energy. Work, in this context, might be the distance a vehicle can travel at a given speed, the elevation to which water can be pumped above a lake, or the weight of freight that can be moved by a locomotive. The energy input would simply be the amount of fuel consumed to do the work. But the term *efficiency* is hardly limited to the world of engineering. An efficient writer might write lots of words per minute, an efficient painter might cover more square feet of wall per minute, an efficient cook might fry more eggs during the breakfast hour, and an efficient distillery might produce more whisky bottles per year.

It's obvious, of course, that efficiency may not be the most valuable measure of a particular action. The efficient writer might quickly generate lots of words with few useful sentences and no story. The efficient painter will never paint a *Mona Lisa*. The efficient cook might produce platefuls of slop. And the efficient distillery might produce large volumes of awful grog. It could be worse, of course. Imagine an efficient vacation (no time for the water park, kids!), or an efficient opera,[2] or an efficient lover . . .

Economists use the term *efficiency* in other ways. *Economic efficiency* is a general term considering the amount of growth—usually GDP[3]—added to an economy per unit of energy. Economists appear to be able to reveal an environmental miracle when they declare that a nation's economy is generating growth without an increase in energy consumption. This is also called energy intensity or decoupling (see chapter 9). *Pareto efficiency* refers to an economic scenario that makes one individual better off without making another less well off.

A particularly important efficiency concept is the efficiency with which energy is produced. It always takes energy of one sort to make energy of another. Turbines must be built before the energy of the wind can be harnessed, and wells must be drilled before oil can be extracted from the ground. The efficiency of energy production is sometimes called net energy, energy return on investment (EROI), or energy returned on energy invested (EROEI). It might be Watts per Watt or barrels of oil per barrel of oil, but it is always presented as a ratio. A high EROEI means that a given source of energy is easy to tap and put to work: picture a Saudi Arabian oil field from which the black gold flows freely when wells are drilled into it. A low EROEI means that a lot of energy must be spent to recover energy: picture the Alberta tar sands, which must be mined with huge excavators, and then the oil must be squeezed from the sand.

Efficiency has taken on a whole new meaning in recent years, and it has become an even greater virtue. Everyone knows that a Toyota Prius is much more efficient than a Hummer, and so Prius drivers are virtuous and Hummer drivers are jerks. There's no confusion here. If you bought a refrigerator, heater, or dishwasher recently you probably felt virtuous for getting the one with the highest Energy Star rating or a little guilty if you didn't. Perhaps you are among the supervirtuous and have bought a car that runs on biofuels rather than gasoline.

But what if there is more to efficiency than meets the eye and it's not really all it's cracked up to be? First, we seem to be missing out on a few things. I used to love writing letters, for example—well, receiving them, at least—but e-mail became more efficient, and letters became news clips, and now e-mail is giving way to superefficient text messages such as this rich, nuanced message that I recently received from my son: "dad where r u :(pls get me asap luv d :)." I am forced to go to Walmart every now and then because it's the only place I can get a stapler, a DVD, parakeet food, zit cream, and new undies all in one outing. While there, it's even more efficient to pick up the groceries as well—but I'm always disappointed when I cook the pack of water-logged chicken breasts and tasteless tomatoes (oh, and by the way, you should get a second stapler while you're there for when the first one breaks).

We lose some of the beauty of life when we become evermore efficient, that much is obvious, but it's all in a good cause, right? We need to make progress and get things done, and we need to do it as efficiently as possible to save resources.

Are you sure?

What if there is a much bigger problem with our eternal drive for efficiency? What if our guilt-ridden quest for efficiency makes matters worse? What if the virtuous Prius driver consumes more energy than the jerk in the Hummer? What if the highly rated Energy Star refrigerator that Mr. Smug bought burns more energy than your cheapo knockoff? What if the ethanol in your gas tank actually increases your carbon footprint?

Part of the problem with efficiency, perhaps, is merely semantic. When we claim efficiency we imply that we are also conserving—but efficiency and conservation are not the same thing at all. Efficiency can mean doing the same job with fewer resources, which is conservation, but it can also mean using the same resources to do more, which is not.

This idea that we need to avoid conflating efficiency with conservation is one of the key messages of *The Efficiency Trap*. Efficiency is no way to save, as we shall see, but this might not even be its biggest failing. Efficiency can have other effects on dynamic systems that can change them in surprising ways.

Our drive for efficiency has become an imperative in all manner of economic and social systems. In business, the vogue is "lean," and tomes have been written on the virtues of streamlining business models for efficiency. The same is true of public services. How can we teach our children more efficiently? How can we run our hospitals more efficiently? The whole process of globalization, buying cheap Christmas presents from China and offshoring our phone operators to India, is a process of global economic efficiency. The focused drive for efficiency can have all sorts of undesirable effects.

In order to visualize the role of efficiency in the dynamics of complex systems it is helpful to evaluate its effects in natural systems: ecosystems. Many natural systems cycle through stages of expansion, maturation, collapse, and renewal that are ominously reminiscent of those found in boom-and-bust business cycles, cycles of war and peace, and the rise and fall of civilizations. Efficiency is not a sign of stability in these cycles, but a sign of danger. Efficiency may maximize a system's productivity, but it can also strip it of its strength.

The problem of efficiency is a difficult one to discuss because efficiency screams out to the environmentally conscious as necessary and valuable: a moral obligation. As a result, it is logical to expect that the only ones speaking out against efficiency will be antienvironmental apologists, such as industrialists who wish to protect their right to pollute or the politicians who kowtow to them. I am an environmentalist and no industry apologist, and yet I'd like to show you why efficiency is the worst nightmare of the environmental movement and why new technologies can never save the world.

PART 1

THE EFFICIENCY PARADOX

CHAPTER 1

THE PETROLEUM INTERVAL

Why Now? Why So Suddenly?

D oes it strike you as odd that this glittering modern world appeared when it did? We humans are a pretty old species and we've been every bit as smart as we are today for hundreds of thousands of years. We've also been experimenting with civilization for a long time. Complex societies arose in various corners of the world thousands of years ago, and yet most of our history had no internal combustion engines, no phones, no planes, no plastics, no computers, and fewer than a billion inhabitants, most of them farmers.

John Maynard Keynes rather crudely designated this great stretch of human history as an "era of scarcity" and called the short, modern era that has followed an "era of abundance." Why did this abundance appear so suddenly? This second era is so incredibly short, so young. It represents barely a ten-thousandth of our history as a species.

Efficiency appears to be a modern idea that applies mainly to machines, particularly those powered by fossil fuels and electricity, but the concept of efficiency is also relevant to the preindustrial era, and it's a logical place to start thinking about how our love affair with efficiency began at the dawn of humanity a million and a half years ago when, in the northern reaches of Africa's Great Rift Valley, some high-achieving primates began tooling around with rocks.

Certain rocks chipped and fractured into a range of different shapes that could be used to do a number of tasks more efficiently, such as impale an animal and cut it into steaks and chops. It took a long time for early humans to figure out that stones could be chipped into intricate shapes and that they could also be hafted onto a wooden or bone shaft. The evolution of tools represents a step-by-step improvement in efficiency, measured as the amount of food calories generated per hunting hour, or per life lost. With simple but efficient hunting tools in place, humans had become the most dangerous predator on the planet, so efficient, in fact, that they could now cause the first major human-induced environmental catastrophes.

The first Americans arrived in the Western Hemisphere on foot by crossing the Bering Land Bridge around fourteen thousand years ago. They carried advanced weapons with them, and they knew how to use them. Humans represented a level of technological efficiency never before seen in the Americas and they spread across the continent like a wildfire across the prairies. A rich diversity of societies settled every corner of the Americas from the Canadian Maritimes to Tierra del Fuego within just a few thousand years, but the efficient hunting tools that powered their colonization also signaled the end of the rich American megafauna. Scores of species fell into extinction at the points of the hunters' spears.

Another major step up the ladder of human progress was agriculture, which further increased the efficiency with which humans could acquire food. Agriculture was probably brought on by the overhunting and overgathering of societies that settled in one place, but, once it was sufficiently established, agriculture enabled the production of food surpluses and had far-reaching impacts on society. People could now live more easily in areas with seasonal climates by generating food surpluses to survive the offseason, and they could provide insurance against tough times when the amount of easily hunted or gathered foods might be limited. Most importantly, food surpluses enabled societies to become increasingly complex and stratified. With a few efficient farmers growing extra food, other members of society were freed up to do other tasks.

Another important impact of efficiency on the advance of civilization was the harvesting of wood resources with stone and then metal tools. The increased efficiency of tools enabled more efficient harvesting, but it could also accelerate depletion, and deforestation has been repeatedly linked with civilization collapse.[1] Remote Easter Island was forested when it was first inhabited by humans around 900 CE, and its wood resources enabled the formation of the thriving society that built and transported the hundreds of iconic, glowering Moai statues that lined its shores. Deforestation brought about the collapse of Easter Island's civilization barely seven hundred years later. The Anasazi built a complex society in the fragile environment of the American Southwest but, once they had consumed all the local pinyon pine and juniper trees and had to trek all the way to the mountains for ponderosa pine, their fate was sealed.

The most dramatic example of forest-dependent rise and fall, albeit the most contentious, is probably ancient Rome, an immense empire that grew to dominate most of Europe, West Asia, and parts of North Africa. The Roman Empire became a huge resource-extracting machine, and its most valuable resource was wood. Wood was consumed in prodigious volumes. It was the base material for construction, including the massive fleets of merchant and military vessels, and it was the energy source for all Roman industry, such as brick making. Colossal public baths required huge volumes of wood to heat water, and massive volumes of wood were roasted into charcoal to smelt metals.

Thousands of historians have speculated on the reasons for the sudden and catastrophic fall of the Western Roman Empire, but the environmental impacts are so clear that they must have played a pivotal role.[2] Rome's population exploded and its empire flourished in complexity through successive developments in efficiency as it tore resources from its land base. Testament to the environmental damage perpetrated by Roman efficiency is the era that followed. While it is often claimed that the fall of Rome was brought about by the superior military tactics of enemies on its periphery, notably the Huns, that seems to be of secondary importance. When the Huns and other "barbarians" finally sacked Rome, they inherited virtually nothing. The end of the Roman Empire did not mark the beginning of another grand empire but the beginning of the Dark Ages. Roman, efficiency had stripped the land of trees, eroded soils, and hollowed out the entire Mediterranean region. The Roman Empire was dead long before Alaric the Visigoth sacked Rome in 416 CE. He just showed up to set ablaze what remained. It took Europe a very long time to recover from the environmental atrocities of the Roman Empire, and it languished in the Dark Ages for half a millennium.

A number of power centers eventually reemerged in Europe, and the Dark Ages gradually gave way to a renaissance. A number of countries emerged as significant powers and colonized foreign lands to extend their spheres of resource extraction. Britain became particularly powerful, dominating the most lucrative colonies. Its economic grip was maintained by the theft of people from Africa to service some of these colonies. Slavery became a government-sanctioned endeavor of brutal efficiency.

As its influence and economic might grew, however, Britain began to strip its land base of resources in its turn. Wood was becoming a particular problem, and Britain was teetering on the edge of the familiar trap that had ended so many civilizations before. Its disappearing forests might have shrunk the British navies and arrested its economic development, but the shortage of wood in this place, at this time, had different consequences. Instead of casting the growing British Empire back into obscurity, wood depletion spurred its embryonic industrial machine to reach for coal. Peasants, unable to afford wood, had used coal for some time, but the nasty, sooty, polluting black rock was considered a poor substitute by most. As coal began to be consumed on a large scale, however, it became clear that it had one enormous advantage over wood: pound for pound, it contained more energy.

The Thermodynamics of Civilization

> Coal in truth stands not beside, but entirely above, all other commodities. It is the material source of the energy of the country—the universal aid—the factor in everything we do. With coal almost any feat is possible or easy; without it we are thrown back into the laborious poverty of early times.
> —William Stanley Jevons, *The Coal Question*[3]

I think most people see the modern world as an era fueled by ingenuity; they attribute the emergence of our high-paced, modern world to hard-working, innovative humans and the technologies they have devised. I don't think this is the case at all. Humans have been ingenious for hundreds of thousands of years.

What has changed is the materials at our disposal.[4] The best minds among the ancient Maya developed complex canal systems, towering pyramids, writing, and a calendar. They built the magnificent Caracol observatory to understand the movements of the celestial bodies. They were really quite brilliant, but they were constrained by an environment with limited resources. The Neolithic Brits carved immense slabs of rock out of the Marlborough quarry and erected the astronomical observatory of Stonehenge. The ancient Greeks built the Parthenon; the Romans built astoundingly impressive coliseums, aqueducts, roads, baths, and cities; and the Mississippians built the city of Cahokia. We have been brilliant and inquisitive for a very long time, and we have left an undeniable archeological record to prove it.

What brought on the sudden emergence of the modern world was not an advance in our ingenuity but the appearance of a new and unexpectedly powerful source of energy. Once we started using it, our ingenuity soon found new ways of putting it to work. The major leap toward the modern world was made with the adoption of coal and the machines devised to harness it. The Industrial Revolution was an era of ingenuity and technology, and of machines and their development, but it was much more fundamentally an era of coal.

We take fossil fuels so much for granted, but it would be a great mistake to underestimate their importance. Fully two-thirds of the world's energy supply comes from coal, oil, and natural gas. Nearly all our advanced technologies are dependent upon fossil fuels. Many of them, from pesticides to paint and plastics, are also synthesized directly from oil. Nitrogen fertilizers are synthesized directly from natural gas. Fossil fuels are the wellspring of our modern world. They might seem to be just some of the many products of the modern age, but, as Jevons said of coal, "they stand not beside, but entirely above, all other commodities. They are the universal aid—the factor in everything we do."

The petroleum interval began in earnest only a century ago with the opening of the Texas oil fields. We sit at its peak today in a world that produces and consumes approximately eighty million barrels of oil per day. It will end about a century from now when we are coaxing the last dregs of oil from shrinking wells. The petroleum interval is a two-hundred year glitch in the history of civilization.[5] Two hundred years is not a long enough historical period to deserve classification as an age or an era, but, as brief as this interval may be, it is one of the most consequential in the history of humankind.

The petroleum interval is a subset of the slightly longer fossil fuel interval, which began about two and a half centuries ago with the first machines fueled by coal. Our coal supplies will outlive our oil supplies by only a few decades and, by the middle of the next century, the last nasty, sooty dregs of the planet's coal will have been scratched from the deepest mines, too—and the fossil fuel era will be over.

The petroleum interval is a fascinating period in the history of civilization. It arrived in the blink of a historian's eye and it will be gone as fast as it came. What will we have to show for it? Will the historians of the future report that we used this once-in-a-lifetime opportunity to set humanity on a path to a harmonious society—or a path to the stars—or will it be recorded as a passing era of wars, environmental destruction, and, well, weirdness?

Fossil fuels did nothing less than change the thermodynamics of civilization. Our species spent hundreds of thousands of years building societies and civilizations from the resources of its immediate environment. A few civilizations expanded beyond their borders to become empires by commandeering the resources of a broader region, but the model remained the same: energy was collected from the environment as it arrived from the sun, the blowing wind, and the flowing water. The dominant sources of energy were food and wood.

None of the great civilizations of antiquity discovered fossil fuels. The Roman Empire might have gone from strength to strength had there been sufficient coal resources in the Mediterranean region, but the coal fields lay to the North, and their power went unnoticed, except by a few peasants. The Sumerian and Mesopotamian civilizations of central Asia rose and fell atop the world's great oil fields. They had no inkling of the immense energy potential that lay beneath their feet—and they were tantalizingly close. The great ark described in the fabulous *Epic of Gilgamesh*, which was etched onto stone plates in primitive cuneiform nearly five thousand years ago (and then plagiarized in the Bible), was waterproofed with pitch—a giveaway to modern geologists that abundant oil lay below.[6]

Huge packets of energy—the remnants of a previous age—had been stored underground. In the case of coal, they were mostly plants that had fixed the energy of the sun into their limbs during the Carboniferous era. In the case of oil and natural gas, they were mostly blooms of microbes that had been deposited by currents into the deep alcoves of ancient tropical oceans. Crushed underground into concentrated packets of hydrocarbon, coal, oil, and natural gas were energy-dense gifts from the sun of an earlier age. No longer was it necessary to harvest the scattered power of the earth from trees, wind, and flowing water. Humans could now tap into the stored energy of an earlier time.

The change was rapid and profound. Machines emerged that could do old jobs much more effectively than before, and then new machines emerged that could do entirely new tasks. Mines could be sunk much deeper and water removed from them more easily. Trains could move goods much faster and in much greater volume than horses, and steam ships began to replace sailing ships. New dyes could be synthesized from coal, and a thousand new industries sprang up. The quality of steel improved dramatically, smelted now from coke rather than charcoal. Coal had changed the world.

And then, in the second half of the nineteenth century, our second great energy inheritance was uncovered beneath the fields of rural Pennsylvania. Oil may have remained a novelty had it not been for the Industrial Revolution (still used only for weatherproofing arks?) but, in this new era of growth and burgeoning industry, it was not long before the power of oil was understood. Here was another source of geologic energy even more energy dense than the last, and it came in a highly versatile, liquid form to boot.

In the early years of the twentieth century, the Industrial Revolution quietly morphed into the petroleum interval and the march of civilization accelerated once more. Cars were invented, and diesel trains replaced steam. Millions of miles of road were built. Electricity proved itself and was soon distributed by massive networks of wires. It was not long before people were routinely flying across the Atlantic and Pacific Oceans by the thousands. The mundane science of chemistry found a new lease on life and spawned plastics. Within a few short decades following the adoption of oil as the energy currency of the world, we had pesticides, iPads, breast implants, Barbie dolls, CAT scans, and polyester undies.

So after a hundred thousand generations of hunting and gathering and a thousand generations of farming, how did we suddenly erect this monumental, fast-paced, global civilization of seven billion in just ten generations? How did we make the transition from scarcity to abundance so quickly, and why did we begin the transition when we did? The explanation is fossil fuels. Coal heralded the Industrial Revolution, and then oil and natural gas followed. The black rock and the black liquid did nothing less than change the thermodynamics of civilization.

The role of energy in the advance of civilization and the growth of economies seems so clear and direct, yet economists have almost never considered energy in their theories. It's a bit of a mystery to me as to why this might be, but perhaps the classical economists were so concerned with the role of labor and the neoclassical economists with the role of capital that they simply overlooked the importance of energy. I guess it makes sense, given that energy has almost never limited economic growth in the developed world. Energy availability, therefore, has simply been ignored in economic calculations, and economists seem to have assumed that abundant energy will always be delivered. Their baseline is one that constantly grows, and since all the existing economic models have been developed during the present era of unlimited energy, they therefore adopt this assumption. Any mention that our energy resources could become limited is met with the assumption of infinite substitutability: should a demanded resource go into decline, the markets will supply an adequate substitute and set an acceptable price. There is no theory that can help us understand how economies will react when energy resources go into decline. This is bad, because that's exactly what is going to happen.

A Double-Edged Sword

The modern world was made possible by unlocking geologic sources of energy. Rather than tapping the energy of the sun as it arrives each day or as it is stored for a few years in living plants, we have been tapping millions of years' worth of sunlight energy that were stored by our long-dead ancestors. Uncorking that energy supply has enabled us to power our latest experiment in civilization to dizzying heights, but it's not only energy that we have uncorked. Those vast deposits of stored carbon are now being returned to the atmosphere as carbon dioxide. Human development has now reached a size and scope that is affecting the climate of the planet. Meanwhile, the fossil fuels that powered the construction of this modern world have powered the destruction of the natural world in even greater measure.

The first Americans sent scores of animals into extinction with stone-tipped spears, but now we have guns. The Easter Islanders and the Romans decimated their forests with axes, but now we have chainsaws. The Maya over-worked and eroded their soils to feed growing populations, but now we are trying to push the land to feed a population of seven billion and counting.

We face a long list of problems. High on the list are deforestation, soil erosion, freshwater depletion, and the collapse of fisheries. Each of these prob-lems has been accelerated by a growing and increasingly insatiable population that wields the power of fossil fuels. These are interesting times, indeed: fossil fuels are double-edged swords.

But even as we are coming to grips with the devastation that our fossil-fuel-powered world is suffering, we are also beginning to worry about the fossil fuels themselves. They are beginning to run out. From an environmental per-spective, this might be a good thing. Since fossil fuels have brought so much devastation to the planet, perhaps the planet will fare better once they are gone. This may be true, but what of ourselves? We have become accustomed to this modern world, trapped in it. How will we survive without our planes, trains, and automobiles? And there are seven billion of us now: can the world feed so many people without fossil fuels?

This book is not the place for a long treatise on energy reserves, a subject that has been discussed in detail elsewhere.[7] There are intense arguments about when peak oil will strike. I'm quite convinced that peak oil is upon us, and it should be noted that even the most optimistic analysts expect the peak to come within two decades. My analysis is that we have already reached peak oil. We may maintain production at roughly today's levels for a few years, but by 2015 it will be clear that the petroleum interval is beginning its inexorable decline. In the broader analysis, and for the purposes of discussing the merits of efficiency, it really makes very little difference whether we mark the oil peak at 2010 or 2030. In a historical sense, peak oil is now—in the first half of the twenty-first century.

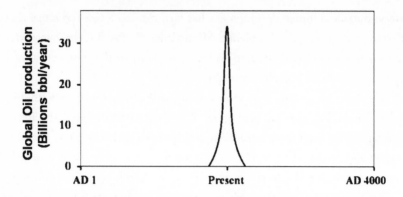

When will global oil production peak? I think it already has, but there are those who think it won't peak for another decade or so. In the big picture it makes no difference. Peak oil is now, in the first half of the twenty-first century, and it is a civilization-level event. *Courtesy of Steve Hallett.*

And it's not just oil. We are being told that we have huge reserves of natural gas, but this is patently absurd: its reserves are severely limited, especially in North America, despite the "wonders" of fracking (more on this later). Reserves in the Middle East and Russia will last for some time, but natural gas is difficult and expensive to move long distances, and shortages will strike many nations soon. The global production peak for natural gas will probably strike around 2025.

The biggest concern for our energy future is probably oil because it is the most versatile of our primary energy sources, but coal is actually our dominant source of energy, and it, too, is coming into shortage. The repeated claims that we have "hundreds of years" of coal reserves are beginning to sound hollow. It should be noted that reserves estimates for British coal were nine hundred years in 1864 and five hundred years at the turn of the century.[8] By 1984 the Brits were predicting approximately ninety years' supply. Really? The coal industry is now little more than an energy afterthought for Britain. But there is much more coal in the United States of America, right? Well yes, there is, but at the turn of the twentieth century the US Geological Survey (USGS) estimated that the nation held a five-thousand-year supply. By 2009 that estimate was down to two hundred years. How did we lose forty-eight-hundred-years' supply of coal? And how sure are we that the new number is anywhere near the mark? It seems highly doubtful that we will be powering our societies with fossil fuels for anywhere near that length of time.[9]

The situation is even more precarious in other coal-dependent economies. The biggest threat of coal depletion is in China. Its massive economy is deeply dependent on coal, but its supplies are depleting rapidly. That China has begun to import large volumes of coal in recent years should be seen as a very worrying sign.[10]

China is deeply coal dependent, but it is assuring everyone that it has vast reserves—so why are so many of the coal bunkers along the Yangtze River empty? *Courtesy of Steve Hallett.*

These are interesting times. The world appears to be both mesmerizingly beautiful and powerful and yet, at the same time, somehow riddled with ailments and vulnerable. It's awfully difficult to see a path into the future that can maintain the beauty while limiting the damage. Awareness of our challenges has grown enormously over the last few decades and we now have a detailed understanding of many them—when considered in isolation, at least. Many groups have mobilized into action and thousands of groups, from small communities to committees of national policy makers and their think tanks, have prescribed a range of actions that might take us forward along a path to the future that will protect our societies and our environments. There is an enormous range of prescriptions and a few dominant themes. A conventional wisdom on sustainable development and environmental protection has emerged.

CHAPTER 2

THE CONVENTIONAL WISDOM

Energy efficiency is a worthy goal. When we exchange old incandescent light bulbs for new LED lights that use a fraction of the electricity and last far longer, we save energy and resources—and that's a good thing. Full stop.

—Richard Heinberg, *The End of Growth: Adapting to Our New Energy Reality*[1]

This agreement on fuel standards represents the single most important step we have taken as a nation to reduce our dependence on foreign oil. Just as cars will go further on a gallon of gas, our economy will go further on a barrel of oil.

—Barack Obama[2]

The Growth Imperative

The first item on the agenda of the conventional-wisdom committee is the growth imperative. It seems that, whatever else may happen, we must maintain our lifestyles, and, to do that, we need to maintain some measure of growth in our economies. It is acknowledged that there are problems, that many of them are serious, and that significant efforts are needed to tackle them, but the big issue seems to be how to "get back on track," or how to "restart our economies."[3] Solutions have been proposed from across the political spectrum.[4]

There is considerable disagreement about the strategy that should be taken to solve our grand challenges, but there is general agreement that we need to continue to grow our economies. It's almost heresy to suggest otherwise. The Right generally believes that wealth is created from the top of the economic pile and trickles down to the rest of us. Everybody is better off, they argue, if we support the ability of the rich to generate wealth for society at large. The Left generally supports a stronger ideal of wealth distribution. They don't trust the wealth to trickle down of its own accord and prefer policies that make sure the less wealthy are protected.

But what if there is no growth? There would be no wealth to either trickle or distribute. None of the mainstream political parties has any answer to this, and so they refuse to even admit the possibility. Growth remains imperative.

The response of politicians to the recent recession has been very revealing. My view, and the view of many ecologists, is that our current economic woes may be the beginning of something much more significant than a temporary or even a "double-dip" recession. We may have met the beginning of a very long decline. The time may have finally come when our insatiable consumption has bumped up against impenetrable ecological limits to growth. Oil production is peaking and the costs of many natural resources and ecological services are rising.[5]

Our politicians have no such fears. The recession is a short-term glitch, they assure us, and once we put the economic train back on the right track we will be back to return to business as usual. The bailout packages of the George W. Bush and Obama administrations pumped trillions of dollars into the economy in an attempt to get it restarted. Bush and Obama hardly wear the same political stripes, but, in this, they were in lockstep.

Our politicians are so afraid of the end of growth that they can't even consider the possibility of letting it happen, and it's easy to understand why. A modern economy trapped in zero growth for a prolonged period is a worrisome prospect, and an economy trapped in a long-term decline is just plain scary. Much of our economic activity is based upon the transference of debt. The offering of a loan of any kind, by any lender, works on the assumption of growth. The lender that provides your mortgage, as just one example, is just as keen as you for the value of your house to rise. The entire real estate industry is predicated on growth. It doesn't matter that nearly every house built in the last three decades is a monstrosity of two-by-fours and chipboard built for cost efficiency that can never hold its real value. But even though it's real value— its ability to fight off the elements without leaking or warping—might be declining rapidly, its financial value needs to rise.

Once the conventional-wisdom committee has reaffirmed the growth imperative, it can move on to the next item on the agenda. Here we find an array of issues that threaten growth. At the top of the list are the looming, twin problems of global climate change and peak oil. How to deal with those? A disconcerting number of committee members deny that these issues even exist, but the majority is coming to accept them as real and dangerous, and as serious threats to the growth imperative. What should the solution be?

Killing Two Birds with One Stone

Although it's hard to pick out any of our crises as more dangerous than another, global climate change and peak oil have special importance, and they are indelibly linked. The extraction and combustion of our geologic inheritance of oil is the dominant cause of global climate change and so the obvious solution is to burn less oil. Oil depletion is about to sink us into a pervasive energy crisis as we pass the production peak, and the obvious solution to this is the same: burn less oil. Simple, then: if we reduce our consumption of oil we are killing two birds with one stone.

But how?

The obvious solution presents itself immediately, and it cuts to the heart of the conventional wisdom: we need to use oil and the other fossil fuels more efficiently.

There is a seductive beauty to the efficiency solution because it would seem that it can not only save energy and reduce carbon dioxide emissions but also generate economic growth. Efficiency might be the way we buy time until renewable energy sources can take up the slack.[6] Efficiency can even help countries compete better against one another economically. Efficiency has become such a lauded goal that is has been called the fifth fuel.

The Fifth Fuel

> We think of energy efficiency as the "fifth fuel," joining coal,
> natural gas, nuclear and renewables as a critical resource needed
> to serve the growing energy needs of the communities we serve.
> —Duke Energy[7]

There is no doubt that our societies are inefficient in many ways, and the way we consume energy is a prime example. Consider the energy lost through the various processes involved in the use of coal resources to cool a house with air conditioning. The coal has to be mined, transported to a power plant, burned to heat water to generate steam, which turns a turbine that generates electricity that passes through the grid to your air compressor, which cools air that flows through your ductwork (and then out of the gap in your window frame). There are a number of critical energy conversions involved in this process—coal to heat, water to steam, moving steam to moving turbine, turbine to electricity, electricity to heat, warm air to cool air—and energy is lost at every turn. Indeed, as much as 5 percent of the energy is even lost as the electricity passes along the wires of the grid. This is a particularly paradoxical example, come to think of it, because while the entire objective of the exercise is to generate cool air, most of the energy is lost as heat (have you ever noticed how hot your air conditioner compressor gets?).

The same basic logic can be applied to any system: consider the flow of energy from a reservoir of crude oil a mile under the Gulf of Mexico to the take off of a jumbo jet at Heathrow, or the flow of energy from the sun, through a sugarcane plant, to an ethanol factory, to a car in Rio de Janeiro. The vast majority of the potential of our primary energy sources is lost somewhere along the way from extraction to application. We generate much more wasted heat than useful work.

The basic idea of energy efficiency is to reduce the losses at any and all stages between source and sink. If we can keep a living room at an acceptable temperature with one ton of coal, where once we needed two, then we have saved a ton of coal by being more efficient. Our efficiency measures, in this case, would be equivalent to a ton of coal—and so efficiency can be thought of as a form of energy, just like coal. Two tons of coal are required to cool your living room: one is delivered as coal, but the other is delivered as efficiency. Efficiency, therefore, is the fifth fuel.

The seductive beauty of framing efficiency as a fifth fuel is that it seems to offer a relatively painless way of making valuable changes. We would like to be able to make drastic reductions in greenhouse gas emissions without threatening economic growth, and here is an invisible, emissions-free fuel that can deliver. The other solutions are just too painful. Beg all you want for direct reductions in energy use, but the economists will warn of devastating consequences to the economy and no politician will be able to support it.

The idea of efficiency as the fifth fuel has gained significant traction, and it is probably the mechanism most commonly proposed for reducing energy dependence while reducing greenhouse gas emissions without going broke in the process. Most of our leading environmental authors, such as Lester Brown of the Earthwatch Institute, Bill McKibben, and Richard Heinberg of the Post Carbon Institute stress the importance of efficiency.[8] This is most definitely the mainstream view and the conventional wisdom. A number of detailed reports explain not only that efficiency represents a fifth fuel, but just how much of this resource exists, what policies should be enacted by government to promote its use, and how markets might be developed to trade it as a commodity. The fifth-fuel concept has been presented in a range of different ways. Armory Lovins of the Rocky Mountain Institute has promoted the terms *negenergy* and *negawatts* as another way of stressing the idea that energy saved is energy made.[9]

The fifth-fuel concept seems thoroughly reasonable. It screams out to the environmentally conscious as a thing that they, as individuals, can be getting on with. It is not too scary to industrialists: they can afford to get on board, too, and it can help their image in the process. Here is the most valuable way forward. It is relatively painless, and it will make the world a better place.

Well, that's the conventional wisdom, at least.

Feeding the Nine Billion

> Corn is an efficient way to get energy calories off the land and
> soybeans are an efficient way of getting protein off the land, so
> we've designed a food system that produces a lot of cheap corn
> and soybeans resulting in a lot of cheap fast food.
>
> —Michael Pollan, interview in the movie *Food, Inc.*[10]

The conventional-wisdom committee is all on board with the energy plan: increase efficiency as a way of generating economic growth while protecting the environment. Check. The next item on the agenda is food.

The problem is not difficult to understand: food production is being pushed to the limit and the global population, currently seven billion,[11] is expected to reach nine billion by midcentury.[12] The conventional wisdom is pretty easy to understand, too. To address this problem we need to produce more food for those two billion more mouths. In order to do this, we need to raise yields. In other words, we need to convert sunlight, water, and nutrients into calories more efficiently.

Everybody acknowledges that this is not going to be easy—but we've done it before, so surely we can do it again, right? There have been repeated warnings about worldwide food shortages, but they have never come to fruition. Thomas Malthus was the first to cry wolf at the turn of the nineteenth century, but he failed to see that the limits to food production would be lifted by row cropping, machinery, and advanced fertilizer (the Industrial Revolution, in a sense). Paul Ehrlich cried wolf next. Poor old Paul was so thoroughly lambasted for predicting famine in South Asia in the 1970s when it didn't happen. Yields boomed through the years of the so-called Green Revolution as new high-yielding varieties were developed for a world flush with fertilizer and new irrigation technologies.[13]

So here we are again, and we need to take the next step. It's a daunting challenge, to be sure. The global population is more than double what it was during the Green Revolution, and it's also much more urbanized. We've lost a lot of land since then as well, some of it buried under concrete, some of it degraded, salinized, eroded, or claimed by desert. But the consensus view seems to be that we can manage again. These Malthusians are always wrong—and who likes a pessimist? What we need to do is double down, focus on efficiency and productivity, and raise those yields. Don't tell me it can't be done: that's called crying wolf.

The agricultural efficiency machine is certainly impressive, and modern agriculture is no peasant activity. It is a high-tech enterprise, and it is supported by impressive scientists. In the United Sates, agriculture enjoys support from public institutions such as the land-grant universities and the US Department of Agriculture (USDA). Lots of smart people are applying advanced science, including biotechnology, to the advancement of agriculture. Private research is carried out by a number of agribusinesses. These folks are hell-bent on making agriculture bigger, better, and more efficient so that yields can be raised once more.

The success of the Green Revolution is generally hailed as a triumph of plant breeding, and the new crop varieties of the 1960s were impressive. Breeders developed crop varieties—wheat and rice, in particular—that allocated much more of their energy to developing and filling grains rather than growing shoots and leaves. Dwarf rice is a case in point. Rice plants were bred that produced stunted offspring with extra-large panicles of seeds, and since it is the seeds not the stems that we harvest, this is a very big deal. The additional advantage was the way these plants responded to fertilizer. Adding extra fertilizer to traditional rice varieties would simply encourage them to grow bigger and taller and fall over, but these new varieties used the fertilizer to produce still more seed. The combination of plant breeding and cheap fertilizer boosted yields tremendously, often doubling or tripling them.

Plant breeding has had no massive breakthroughs since the Green Revolution, but yields have continued to increase as plants have been continually tweaked and the science of plant breeding has become more sophisticated. Plant breeding can be assisted by molecular techniques that enable breeders to focus in on specific traits that promote high yields, and now genetic modification can incorporate novel traits into crop genomes. A plant can't be made to do just anything, but it can be made to do many things you wouldn't expect. Crops can now carry their own insecticides, and they can be made resistant to herbicides by being endowed with genes from bacteria.[14]

The possibilities of crop improvement with a combination of advanced breeding and genetic engineering appear to be almost limitless. Researchers are working on crops to improve their drought tolerance, salt tolerance, and heat tolerance (all of which would seem to be useful in the drier, saltier, and hotter world to come). The general consensus is that the researchers will, as they always have, succeed.

Efficiency is the solution most often proposed to solve our growing global food-supply problem. As with energy efficiency, it is almost heretical to question this, and another case of crying wolf. We can and must get over this latest food-demand hump, and the wolf, we are told—because of our drive, determination, and brilliance—will not come. But that's not how I remember the old fable at all. The fable is told different ways by different people: they might give a version that is short and sweet, or they might drag it out for hours, but one part of the story is always the same: the wolf always shows up in the end. And in any case, what happens if we do manage to feed the nine billion by midcentury? Will we then be asked to feed twelve?

Greenwashing

Awareness of our environmental dilemmas has grown considerably in recent years. Most of the issues are now covered in the media and taught in our schools and colleges. The desire to avoid and ameliorate environmental problems has also grown considerably, and most right-minded people are concerned and willing to contribute where they can. There are obvious exceptions, of course, and some horrendously biased media, such as Fox News, that are only suitable for viewing by sheeple, not people, but the awareness that modern society is causing significant environmental harm is now strong, and many people want to do their part. The recognition of our looming energy crisis is somewhat less broadly acknowledged, but awareness is growing there, also.

Take global climate change, for example. Most people now understand the link between greenhouse gas emissions and global climate change, and most of those would like to reduce their carbon footprint if they could.[15] The changes most people make are rather modest, however, and the translation of awareness into action has been poor. Even as our awareness that carbon dioxide emissions cause global climate change has grown, so have the emissions themselves.

There are a number of difficulties faced by well-meaning citizens. First, there are pervasive institutional barriers erected by corporations and politicians that actively resist change. They have a vested interest in maintaining the status quo and economic growth, and policies that would curtail growth are extremely difficult to pursue at the institutional level. Probably even more pervasive is the way individuals become trapped in a high-consumption lifestyle. It's awfully difficult to get out of the rat race once you're in. If you have a mortgage, you probably want the value of your house to rise. You probably want the price of gasoline to stay low. If you have kids in college, you probably want a raise. The changes you are willing and able to make, then, are within this basic framework, and even the most environmentally aware citizens are looking for changes that can help within their own constraints.

This is the seductive logic of efficiency. It appears to offer a way of contributing to long-term environmental sustainability within the bounds of protecting one's short-term financial sustainability.

The two quotes at the head of this chapter are revealing. Richard Heinberg tells us, in no uncertain terms, that "efficiency is a worthy goal," because ". . . we save energy and resources, and that's a good thing. Full stop." It can be a worthy goal, but there is definitely no full stop. There is much more to be said about efficiency. President Barack Obama unwittingly gives us a huge clue that there might be a problem when he lauds the virtues of efficiency. What is the value of efficiency in his view? It is not that energy is saved, or that emissions are reduced. What is much more important, it seems, is that "cars can go further on a gallon of gas and the economy can go further on a barrel of oil." I single out Richard Heinberg and Barack Obama not because I am opposed to them. Quite the contrary. Heinberg is one of my favorite authors, and he has illuminated the perils of the modern world as much as any other author in the last decade. That he should make this mistake, then, indicates to me that very few are immune. Barack Obama is, in my view, an outstanding president and person, but he's the driver of a runaway train. He is blamed from both Right and Left for any manner of misstep, but he may have done more than anyone else could have to maintain a relatively cohesive society through such volatile times. His mind-set, though, while it is much more proenvironment than that of his opposition, cannot see past the economy. Efficiency is an activity he can get behind and support because it appears to protect the environment and pleases the environmental lobby. What makes it even more politically appealing is that it supports the growth imperative.

Many people have no fear for the future and are pushing ahead blindly. Others have their eyes wide open but are fixed like deer in the headlights, not knowing how to help. The biggest problem, however, is with the large and growing group of people who are fully aware of our worsening predicament and are doing what they can. These are the people who can change the world, but right now most of them are doing the wrong thing. We have been seduced by the conventional wisdom that tells us that fossil fuels can be saved if we consume them more efficiently. They cannot. It also tells us that global climate change can be mitigated by efficiency. It cannot. We also want to believe that new technologies will come along and save the day. They will not.

So are we stuck? If efficiency is a trap, is there nothing we can do? Please read on. There are lots of ways to make the world a better place, but the first step is to understand the efficiency trap and break free of the conventional wisdom.

CHAPTER 3
THE UNCONVENTIONAL WISDOM

Whenever you find yourself on the side of the majority, it is
time to pause and reflect.

—Mark Twain[1]

The Coal Question

It is very commonly urged, that the failing supply of coal will
be met by new modes of using it efficiently and economically
. . . [that] the coal thus saved would be, for the most part, laid
up for the use of posterity. It is wholly a confusion of ideas to
suppose that the economical use of fuel is equivalent to a
diminished consumption. The very contrary is the truth.

—William Stanley Jevons, *The Coal Question*[2]

The unconventional wisdom is not at all new. It was first presented by
William Stanley Jevons in 1865 in his seminal work *The Coal Question*.
Britain was the world's dominant political and economic power at
that time, and it was the world's biggest producer and consumer of coal. The
importance of coal to Britain's imperial might was not overlooked, and even
though the consensus view was that Britain's coal would last for centuries,
a few analysts were beginning to worry about its possible depletion. They
argued that action was needed to extend the useful life of British coal and that,
to save coal, the country should invest in science and technology and develop

engines that burned coal more efficiently. Jevons laid out his counterargument in *The Coal Question*.

Jevons predicted that British coal production would peak soon after the turn of the twentieth century and then go into decline. This, he felt, would bring economic catastrophe to the British Empire. His date for peak coal was spot on, and the British Empire was eventually reduced to the Rock of Gibraltar and the Falkland Islands.[3] (He also predicted, presumably to roars of laughter, that the prodigious energy resources of the United States would propel it to a position of global economic dominance).

Jevons also completely disagreed with the conventional wisdom that increased efficiency would prolong the life of coal. As he said, again and again, "the very contrary is the truth." He recognized that the efficient use of coal could never reduce its consumption. Instead, efficiency would act as a force that would, as it always had, increase its consumption.

Jevons plotted the efficiency of steam pumps through time. The first steam pumps, devised by Newcomen and Savery to pump water out of coal mines, consumed approximately 120 pounds of coal to raise a million gallons of water by a foot.[4] The next generation of steam pumps, following the efficiency improvements developed by James Watt, required only half the amount of coal to do the same job. By 1830, following the development of the high-pressure steam engine and the double-cylinder engine, the volume of coal required had been slashed a further four times. Had the amount of coal consumed in Britain declined with these advances in technology? No, it had grown almost exponentially. Steam pumps were now installed in all coal mines and they enabled mines to operate at much greater depths.

George Stephenson started out as a laborer in the coal mines of Wylam, England. Although illiterate,[5] he proved talented at tinkering with the mines' steam engines—and he also tinkered at home. In 1814 he tested his new invention, a prototype steam locomotive, which he called the "Blucher." It successfully hauled thirty tons of coal uphill along a rickety four-mile track. He went on to design and build the "Rocket," the first genuinely impressive steam train, and the rest, as they say, is history. As more and more efficient steam trains were developed, railways spread across Britain like spiderwebs.

Steam engines were built for all manner of other contraptions, such as looms and ships, and further efficiencies emerged. By 1900, steam engines required only a thirtieth the amount of coal to do what Newcomen's engine could do—a staggering increase in efficiency.

But doing the same work with a thirtieth the amount of coal is not what the new, efficient engines did at all. There were now millions of engines doing much, much more, and they were churning through mine after mine full of coal to do it. Efficiently, yes: but with prodigious levels of consumption. Jevons was adamant that the solution to the dilemma of coal depletion was not further increases in efficiency. "The very contrary," he insisted, was "the truth." Replace the word *coal* with the word *oil* and you can hear Jevons's words echoing down through the ages.

Another major innovation of the Industrial Revolution was the Bessemer process, a much more efficient process for the removal of impurities from pig iron and the production of high-quality steel. The Bessemer process reduced the cost of steel production by nearly ten times and reduced the number of workers required. Was this an efficiency saving that enabled the steel industry to reduce its demand for energy? Of course not. As Nathan Rosenberg writes in "Energy Efficient Technologies: Past, Present, and Future Perspectives," "The Bessemer process brought with it a dramatic reduction in the fuel cost of steel-making. Indeed, it was this fuel-saving innovation that essentially transformed an iron industry into an iron and steel industry. . . . The Bessemer process was one of the greatest fuel-saving innovations in the history of metallurgy . . . it's ultimate effect was to increase, and not reduce, the demand for fuel."[6]

Cheap, high quality steel, alongside efficient steam engines, enabled the expansion of railroads and steam shipping lines, the construction of steel bridges and eventually skyscrapers. Efficiency gains through the Industrial Revolution were the innovations that enabled Western societies to delve deeper and deeper into their supplies of coal and build the modern world. Efficiency will not reduce our consumption of fossil fuels and it will not reduce our emissions of carbon dioxide.

Conservation, Rebound, and Backfire

The wise words of William Stanley Jevons, and his enigmatic *Coal Question*, were more or less forgotten until a few researchers began asking similar questions during the oil-embargo years of the 1970s, when oil supplies were tight. The old calls for efficiency reappeared: for oil now, rather than coal. Daniel Khazzoom, studying the economics of energy-efficient appliances in the United States, and Leonard Brookes, analyzing propositions for the reduction of greenhouse gas emissions in the United Kingdom, came to conclusions similar to those of Jevons. Harry Saunders dubbed their findings the Khazzoom-Brookes Postulate.[7]

Khazzoom and Brookes rekindled the debate about efficiency by once again asking the key questions that had been ignored by advocates of conservation by efficiency and reminding us what Jevons had taught us a century before. Conservation is not the same thing as efficiency. Efficiency represents an energy saving only if you consider the replacement of one machine with another that does the same job. But what happens to the energy that is saved by the new, more efficient technology? What happens to energy prices when a more efficient technology comes along? And what are the longer-term consequences of having the more efficient technology?

There are a number of answers to all these questions, but the first effect is straightforward. Increased efficiency reduces the pressure on supplies, and so its first impact is on price. Lowering prices tends to encourage greater use, and so some of the energy savings are returned. The term for this is *rebound*. Energy may still have been saved, overall, but not as much as was expected.

If a car owner trades in his luxury car for a compact, he can easily calculate the fuel savings he ought to make by comparing the improved efficiency, in miles per gallon, of the new car over the old one. Rebound occurs if he drives the new, more efficient car slightly more often, or on longer trips—and this is, indeed, the norm.[8]

Evidence for rebound effects such as these has been sought by a number of researchers. The results are rather variable and depend on the type of analysis performed and the scope of factors included in the analysis, but they have been well summarized by Horace Herring and Steve Sorrell,[9] who found an average rebound from car efficiency, measured across a number of countries, to be in the order of about 20 percent. That is, about a fifth of the fuel savings from an increase in fuel efficiency were given back by increased vehicle use.[10] A similar rebound, also around 20 percent, was found with household heating.[11] People switching to more efficient household heating systems tend to keep the house a little warmer than they did before.

This rebound effect, where efficiency savings are returned by using the improved technology more, is relatively easily understood and quantified. The efficiency gain saves money or time as well as fuel, or makes the technology more effective, and so the technology tends to be used more. This type of rebound is called *direct rebound* because the increased consumption comes from an equivalent, upgraded technology used in the same way as the old one.

Direct rebound is not that big of a deal because the driving habits of an individual who acquires a more efficient car change only modestly, and fuel has still been saved. The saving is 20 percent less than expected, but that still means it's a full 80 percent of what was expected. That's still a significant saving, so what's the problem?

The seductive problem of efficiency really begins with the indirect effects of efficiency, and its importance becomes clear only when we consider populations of people rather than individuals. A little extra driving by one person is not a big deal, but when everybody drives a little more, a lot can change.

If cars are more affordable, thanks to an efficiency increase, then more of them are put on the roads, and if everybody drives a little more, more roads will be needed. Road construction is an energy-intensive activity and so increased road construction can become an indirect rebound effect of increased fuel efficiency in cars. More cars need more roads, but they also need more tires, more gas stations, and more steel. More steel requires more iron ore. More iron ore mining requires more fuel, and the picture becomes much more interesting when we consider these broader systems. The problem is not that we give back a paltry 20 percent of the energy we thought efficiency would save. The problem is that, since efficiency makes things cheaper and better, it drives us down the pathway of progress. This is not a bad thing per se, but we shouldn't fool ourselves that efficiency saves energy, because it does the opposite.

More efficient cars need more roads that support more cars. A lovely, calming place to contemplate this is one of the eight lanes of the Dan Ryan Parking Lot, with its great view of Chicago. *Courtesy of Steve Hallett.*

Indirect rebound is extremely difficult to quantify. The efficiency improvements in cars seep through into increased demand not only for cars, but also for roads, tires, steel, and many other things—but how much of the increased demand in these other industries can be traced back to the efficiency improvement in the car? More roads would be needed by a growing population even if vehicle efficiency did not improve. It's really difficult to measure these impacts, but imagine what our road systems would look like today if the most efficient car we had on them was still the Model T Ford. Roads would be smaller and fewer, and there would be very few complaints that a speed limit of sixty-five miles per hour is restrictive, and our consumption of gasoline would be much less.

Despite the difficulty in quantifying them, it is clear that efficiency improvements can cause massive increases in consumption, and when we consider the direct and indirect rebounds together, they frequently exceed

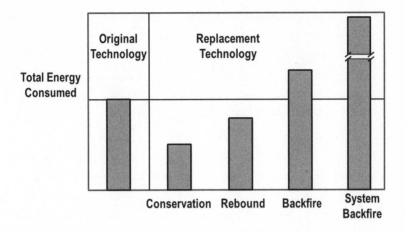

Conservation and efficiency are not the same things. Seldom do we conserve the energy expected with an efficiency improvement. We experience rebound when we use the new technology more. Rebound can become backfire when we do this, but the big effects are on systems. A new, more efficient technology can have impacts throughout the economy. *Courtesy of Steve Hallett.*

---

a rebound of 100 percent. This means that the rebound is greater than the savings predicted in the first place, and the efficiency improvement has caused an increase in consumption. This is called *backfire*.

The efficiency improvements over the century-long history of the automobile have driven massive increases in the consumption of gasoline—a colossal backfire. Why is it that we expect further increases in the efficiency of cars to reduce our demand for oil and reduce our carbon dioxide emissions in the future?

Rebound and backfire are not limited to transportation systems. In fact, they are hardly limited at all. In heating your home you may invest in insulation to reduce heat losses in the winter—a seemingly obvious example of energy conservation but, even in this benign example, there is likely to be rebound. People with well-insulated homes and lower heating bills are likely to keep their houses a little warmer. Cheaper and more efficient insulation enables us to build bigger houses with cheaper materials. A more efficient furnace would appear to conserve energy but here again, efficiency improvements in furnaces tend to make it more cost efficient to build bigger houses.

Better insulation, double glazing, and a better furnace might be just what it takes to convince you that it makes sense to build an addition. We now live in monstrous McMansions that use prodigious amounts of energy. If you really want to save energy, don't pretend you are contributing by getting a more efficient furnace. Sell the McMansion and move into a small stone cottage. Yes, you should then also get an efficient heating system for the cottage, which you'll find effective and cheap to run.

Or consider TV sets. The old tube TVs were very inefficient, but they were generally rather modest in size because of their cost and the amount of space they took up. Sure enough, the energy efficiency of a twenty-inch TV set improved by 75 percent between the early 1970s and the mid-1980s,[12] but did we consume less electricity watching TV during that time? No, we bought more, and bigger, TVs. Enter the LCD screens that are cheaper to run and take up much less space. But what do we do with this new efficiency? We replace the old TV with a much bigger one—often a much, much bigger one. We can also now put one in another room, in the bathroom, or in the car. Heck, the big screens are great for playing Xbox or Wii, so why not devote one to the kids for playing games? Or get an extra one for playing movies? A more efficient screen has been no solution to energy conservation. If you want conservation, turn it off and do some gardening.

What about airlines? More efficient planes carry more passengers and save fuel? No. They result in more passengers, more trips, more planes, and more runways.

What about refrigerators? They are getting more and more efficient and saving more and more energy all over the place. Refrigerators are now saving energy in trucks, shipping containers, and airline holds. They are saving energy by the terawatt in supermarkets all over the world. It's a classic efficiency trap. Inefficient refrigerators could not save all this energy because they couldn't do the job at all. There would be far fewer of them.

What efficiency conserves with one hand it consumes with both, resulting in consumption where conservation was the goal.

Efficiency is hailed as the savior of the modern world because it can save both the planet and our way of life. It is the principle goal of environmentalists and industrialists alike. It has convinced environmentalists that they can save the world in a pain-free way, but efficiency is a trap, and a very seductive one, but this is only the beginning of the impacts of efficiency. There are other, less intuitive, but much more pervasive impacts because efficiency does not only spur consumption where it promises conservation, it can also overburden, simplify, and weaken complex systems.

Efficiency and Progress

> Efficiency quite literally drives growth forwards. By reducing labour (and resource) inputs, efficiency brings down the cost of goods over time. This has the effect of stimulating demand and promoting growth. Far from acting to reduce the throughput of goods, technological progress serves to increase production output by reducing factor costs.
>
> —Tim Jackson, *Prosperity without Growth:*
> *Economics for a Finite Planet*[13]

The advance of civilization has occurred in fits and starts characterized by long periods of relative sameness, periods of rapid development, and sudden collapses. Rapid change can come when a new source of energy is discovered and consolidated through incremental improvements in efficiency. The first major sources of energy were agricultural surpluses and wood. Neither is particularly compact or versatile by today's standards, but they provided the energy upon which many empires were built. Along the way, people figured out how to capture the wind in sails and windmills, tap into the power of flowing water, and harness livestock. We have constantly increased the efficiency with which we deploy all these technologies, but while efficiency can make a base technology go much further, it cannot transform it into something else. This is why the explosion of the human experiment did not really begin until the eighteenth century; it was not until then that civilization reached for coal. The discovery of the power of coal transformed the thermodynamics of civilization, and improvements in the efficiency with which coal was used spawned the Industrial Revolution. The power of coal led to the discovery of oil, and the efficient exploitation of oil led to the modern world.

Energy transitions sit at the heart of the rise and fall of civilizations. Discovery creates the opportunity, efficiency builds the civilization, and depletion presages its collapse. Efficiency plays a central role because it is the mechanism by which the power of energy, resources is put to economic use. The transition from wood to coal represented a pivotal moment in the history of civilization because, pound for pound, coal contained more energy than wood and could do more work: but not much more. On average, coal has an energy density about one and a half times that of wood. The transition was slow at first because wood was being used efficiently and coal was an unknown quantity. It was only when coal was adopted on a large scale that its true power became clear. Recall that the efficiency gains in the steam engine through the eighteenth and nineteenth centuries were in the order of thirty times.

The same is true of oil, which has a little less than twice the energy density of coal. Oil was also adopted slowly at first because it took some time to figure out how to produce and refine it efficiently. Once these technical issues were solved, however, its adoption increased rapidly and soon delivered gasoline-powered cars, converted the railroads from coal and wood to diesel, and took to the skies with aviation fuel and planes. Its adoption has increased every year in the increasingly efficient machines that consume it.

Robert Ayres and Benjamin Warr[14] analyzed the efficiency of electricity generation and distribution over the twentieth century and found that it increased from less than 4 percent to nearly 35 percent.[15] That means that, in 1900, 96 percent of the energy capacity of coal was disappearing as heat and only 4 percent was actually making electrical machines go. The efficiency of energy conversion in transportation rose from 3 percent to nearly 14 percent during that time; that of high process heat generation, such as for the steel industry, rose from 7 percent to 25 percent, and the efficiency of space heating rose from a pathetic quarter of 1 percent to 3 percent. Across the board, this means that the efficiency of energy conversion in primary industries nationwide has risen by roughly an order of magnitude. We have not only increased the volumes of fossil fuels that we have produced to build the modern world. We have also increased the efficiency with which we put them to work.

Neoclassical economics seems to have overlooked the true source of progress and economic growth. Wealth seems to spring up magically when labor, innovation, and capital are put together, but real wealth comes from energy and resources that must first be drawn from the environment. Efficiency drives progress, but it also drives consumption. Most economists seem to be completely unable to grasp the predicament we face.

Efficiency with Declining Supplies

There is an apparent contradiction that needs to be cleaned up at this point. I have argued that efficiency leads not to conservation, but to increased consumption. Meanwhile, in this book's first chapter I argued that global oil production is about to go into irreversible decline. These two things—increased consumption and irreversible decline in consumption—cannot both happen at the same time. So which is it?

In fact, there is no contradiction at all. The two factors will play off against one another. Global oil production will begin to decline very soon because reserves are beginning to go into depletion. There is nothing we can do about this—but it will not be for lack of trying. The demand-driven effect of increased efficiency will simply be swamped by supply-driven depletion.

Disaster struck the other day. The world was coming to an end with wailing and gnashing of teeth. "Dad, Dad! The Nutella jar is empty!" Well, as long as democracy governs and freedom reigns, my children shall have their Nutella. They shall not go without their chocolate spread with a delightful hint of hazelnut. Not on my watch.

Thankfully, an empty Nutella jar is not just a lack-of-Nutella problem. It's also an efficiency problem. In fact, the production of Nutella (I know . . . but if oil is "produced" from a well, then Nutella is "produced" from a jar, goshdarnit!) is a nice model of efficiency processes. Buy a new jar of Nutella, put it in front of the kids, and observe. It goes something like this . . .

When the little foil cover is peeled away, the Nutella sits right up to the rim of the jar like an unblemished sea of chocolatey perfection. In go the fingers. There's really no need for any technological intervention here: spoon or knife unnecessary. It seems like the Nutella will last forever. It comes out so easily, and the prodigious level of waste (mostly over the face) is no big deal. But sticky fingers are only so long, and Nutella production will go into decline without an efficiency improvement. A spoon will do, and with just a little more effort than before, the Nutella can continue to be produced in gobs. A few spoonfuls later, though, it starts to take a little more energy to stay in business. The most easily accessible part of the Nutella reservoir has been depleted and the spoon must spend more time scraping the edges. Eventually, the poor Nutella driller will be expending lots of energy, scratching away as efficiently as possible at the inside of the jar, and yet only the smallest speck of the brown gold will be produced. It's now that the sirens go off: "Dad, Dad!"

You have three options. Option one: "Sorry for your luck, kids." (Not advised.) Option two: go to the store and hope they have another jar. Option three: go one last efficiency step. Set the Nutella jar on the chopping board, grab the meat cleaver (disclaimer: this works only with the plastic jars), and chop the jar in half. I guarantee there's another face full in there. Lick it clean. OK, now it's empty. The kids will not only be impressed; they will also be too frightened of you to ask for more.

You can push the efficiency of Nutella production to the limit with a meat cleaver so that you can drag out one last, unexpected sandwich. But the amount of Nutella in the jar is finite, and even the cutest smiles can't last forever. What's more, increasing efficiency will bring a much more sudden and sticky end. *Courtesy of Steve Hallett.*

Efficiency drives consumption and progress, then it maintains consumption, and then it drives the greatest possible consumption despite depletion. In the end, the highest possible level of efficiency is required to drive the last scraps of a resource into the ditch. Efficiency keeps trying and trying, and it pushes production right to the edge. But some things do actually run out.

Expect to hear this soon, when we pass peak oil: "Oil consumption is declining thanks to greater efficiency." It will seem completely reasonable because efficiency will continue to increase as oil consumption declines, but it won't be true. The correct translation will be: "Oil consumption is declining *despite* greater efficiency." It will become much harder to see the correlation between efficiency and growth. There will be ever-increasing efficiency but, from now on, it will also come with shrinkage.

Consider what happened to the British coal industry. Coal production peaked in the first half of the twentieth century and went into decline (just as Jevons had predicted), but the increases in efficiency did not stop. Larger and more powerful machinery was applied to the task of digging poorer and poorer quality coal out of deeper and deeper mines, and the coal industry stayed alive despite the worsening depletion. It was not until the late 1950s that coal mining in Britain finally began to fall apart. When the problems started, however, they became serious very quickly, and coal production plummeted by nearly half in just two decades. The British government tried to stimulate the coal-mining industry, but pits were closing. The miners' strike of 1984 became the dramatic and often-violent standoff between Margaret Thatcher's conservative government and Arthur Scargill's National Union of Mineworkers, and it marked the end of the power of British coal. British coal once powered an empire, and it was believed that reserves would last for hundreds of years. It was produced more and more efficiently right until the end, but efficiency masked the underlying problem of depletion. When the industry began to fail, it collapsed.

The oil industry is beginning to look like the British coal industry did in the first half of the century. We are pulling out all the efficiency stops with fracking, deepwater wells, seawater injection, and steam-assisted recovery from tar sands because the easy oil is gone. The tougher oil needs to be extracted with greater and greater efficiency but, even as our efficiency improves, and our ability to drag oil out of tougher and tougher places improves, production will go into decline. All efficiency will do is mask the underlying problem, keep the industry running as long as possible, and push us over the cliff more quickly; and soon.

We will be driving extremely fuel-efficient cars in two decades. Our household economics will not only suggest it, as they are doing now, but will demand it. Right now it's a little painful when gasoline prices spike above four dollars per gallon. This has happened a couple of times already, and people are switching to slightly more efficient cars. It's not yet prohibitive to hang onto the old gas-guzzler, but it will become increasingly so. As the oil noose tightens and gasoline prices rise, the economic pressure will increase, and fuel efficiency will rise to match. Efficiency improvements in an era of declining supplies cannot, by definition, increase supply, but they will keep consumption as high as possible.

Imagine that your buddy Jack, down the street, hangs on to his mustard-yellow Hummer (why are Hummers always painted in Ferrari colors?) until the very end. He still has the darned Hummer after all this time. He tinkers with it in his garage. He loves the thing, but I'll bet he hardly ever drives it: it would cost a fortune to fill the tank. He's not the one who is still pushing the envelope of oil consumption, you are. You are now nipping to the local Smal-Mart in your Toyota Weenie that gets a hundred miles per gallon into the wind. The efficiency of the Toyota Weenie (and the Chevy Dolt) has enabled you and millions of other drivers to stay on the gasoline wagon, and it has enabled the gasoline wagon to keep rolling, too.[16]

Efficiency drives consumption but it is not always obvious when consumption must decline because of other, inalienable limits. But efficiency is there, pushing those limits, nonetheless.

PART 2

EFFICIENCY TRAPS

Efficiency, often little more than cost reduction for the boss, has become the holy grail for performance measures in the modern world: the rallying cry of a quasi-moral crusade. The questions "efficiency for whom?" "Efficiency at what cost?" and "Efficiency over what time?" are seldom asked. As a source of light, a normal bulb, for example, is surprisingly inefficient.
Yet as a heater it is very efficient—especially in a sealed box.

—Andrew Price, *Slow-Tech:*
Manifesto for an Over-Wound World[1]

Ned, Gustavus, Henry, and Fred

Ned Ludd is not a fully established historical figure, but he was probably a weaver called Edward Ludlum who lived in Anstey in rural England in the early years of the Industrial Revolution. Industrialization was already changing the English social landscape, particularly in the mines and steel mills, but it was also coming to the textiles industry. Ned Ludd became one of the first people to "rage against the machines" when he smashed the steam-powered looms at a local mill in 1811.[2]

Ned Ludd was ahead of his time, but discontent with the machines of the Industrial Revolution gradually grew across Britain and sabotage became increasingly common. When the authorities interrogated saboteurs they were repeatedly told, "Ned Ludd did it," and the saboteurs became known as Luddites. Their movement grew and consolidated with the general objective of preventing machines from taking away the jobs of people.

The term *Luddite* is generally used in a derogatory fashion ("Oh, come on, Steve, get yourself a proper cell phone. You are such a Luddite!"), and it is largely taken as gospel that the Luddites were wrong on all accounts. Not only were the Luddites a bunch of lawbreaking vandals, but they were also completely wrong to fear that the efficient weaving machines would put them out of work. The workforce employed in the textiles industry boomed. What had been a set of quasi-organized cottage industries across rural England became a large-scale industry employing hundreds of thousands. My hometown, Oldham, in North West England, was transformed by the textile industry. Its population swelled as dozens of mechanized cotton mills were erected. It should be obvious, especially if you've ever visited Oldham, that all that glisters is not gold,[3] but the Luddites were certainly wrong in their main premise. Efficiency through mechanization in the textile industry did not reduce the workforce. The machines put hundreds out of work but created jobs for thousands more.

Efficiencies of this kind were happening all over the Western world because the Industrial Revolution was changing the thermodynamics of civilization. Coal made a thousand industries more efficient, reducing costs and generating capital. The new, concentrated form of energy was creating concentrations of industry and wealth. One such change was happening in the American livestock industry.

With cities growing and wealth increasing, especially on the East Coast, there was an increased demand for meat. The practice of raising cattle all over the Midwest and moving live animals across the country to thousands of butchers and small slaughterhouses near the cities was, well, inefficient. Live animals were hard to move, and they could become sick or die in transit. Meanwhile, the railroads were expanding rapidly, and entrepreneurs like Gustavus Swift stumbled upon a great way to streamline this process and make a bunch of money.

Swift's key innovation was the ice-cooled railroad car, but his true "genius" was using it to link the midwestern livestock industry with the East Coast meat markets by creating the meatpacking industry. The meatpacking industry was efficient, horribly efficient, as the nation found out with the 1906 publication of Upton Sinclair's *The Jungle*.[4] Livestock was moved by train into midwestern cities such as Omaha, Nebraska; Green Bay, Wisconsin; and, with the largest facilities of all, Chicago, Illinois, where animals flowed like a living sea into the packing plants. The animals were slaughtered, sliced and diced with great efficiency, packed into refrigerated railroad cars, and sent off.

In response to the huge problem of animal wastes that were now being generated in the cities, Swift and his contemporaries became innovators in a number of associated industries such as glue, soap, and fertilizer production so that they could, as efficiently as possible, profit from the whole animal carcass. The proud refrain of the Swift and Company was, "We use everything but the squeal." And they did.

Sanitary conditions were often disastrous, especially on unofficial shifts, when the government inspectors were away, and any and all animal parts, as well as lingering rodents and their feces to boot, found their way into the meat, especially the sausages. And it was not just the animals that were treated like meat, but the workers, too, as Sinclair wrote: "The carcass hog was scooped out of the vat by machinery . . . between two lines of men each doing a certain single thing as the carcass came to him. One scraped the outside of a leg, another scraped the inside of the same leg. One with a swift stroke to the neck, another with two swift strokes which severed the head. . . . Another made a slit down the body. . . . Looking down this room one saw, creeping slowly, a line of dangling hogs a hundred yards in length, and for every yard was a man working like a demon."

Upton Sinclair may not have been impressed by the meatpacking lords, but Henry Ford was deeply impressed. He explains in his autobiography, *My Life and Work*,[5] that it was a visit to a Chicago slaughterhouse that opened his eyes to the virtues of employing a moving conveyor system and fixed work stations in industrial applications. Ford saw how this superefficient animal processing could be applied to the manufacturing of his Model T, and so the animal disassembly line inspired the automobile assembly line.

In 1908, before his assembly lines were in place, Henry Ford had made just over ten thousand Model T Fords at a cost of $850 each. In 1913 he made three hundred thousand at a cost of $360, and in 1924 he made two million Model Ts for just $290 each. Before Ford's assembly line it took days to assemble a car. With the assembly line a Model T could be put together in less than two hours and a new vehicle was completed every twenty-four seconds.

The efficiency of the automobile assembly line catalyzed profound changes across the American social landscape, as John Steinbeck points out in his classic *Cannery Row*:

> Someone should write an erudite essay on the moral, physical, and esthetic effect of the Model T Ford on the American nation. Two generations of Americans knew more about the Ford coil than about the clitoris, about the planetary system of gears than the solar system of stars. With the Model T, part of the concept of private property disappeared. Pliers ceased to be privately owned and a tire iron belonged to the last man who had picked it up. Most of the babies of the period were conceived in Model T Fords and not a few were born in them. The theory of the Anglo Saxon home became so warped that it never quite recovered.[6]

America was transformed by the automobile itself, but it was also transformed by the means by which Henry Ford had so rapidly and efficiency put it there, and efficiency became the new mantra of industry.

The first recognized "efficiency expert" was a mechanical engineer called Frederick Winslow Taylor. While working as a machinist at a steelworks in Philadelphia, Taylor concluded that neither the workers nor the machinery were working as hard or as effectively as they could be, and when he rose to a management position in the company he decided to do something about it. He set to studying the factory in a systematic manner, setting tasks to a stopwatch and taking copious notes. He made the factory, and then others like it, much more efficient and productive, and he coined the term "high speed steel." He published his seminal work, *The Principles of Scientific Management*, in 1911.[7] The book was highly influential and set the foundation for a new style of business. Indeed, the newly formed Harvard Business School set a good part of their new curriculum around Taylor's book. Many industrialists followed, and the efficiency movement was born.

CHAPTER 4
ENERGY EFFICIENCY TRAPS

The most significant historical leaps in the development of energy have come from the replacement of less efficient fuels with more efficient ones. In this context, the main efficiency gain is made in energy density, although the versatility of the source (to be transported and refined, for example) is also important. The historical progression from animal feed to wood to coal to oil and natural gas is a natural progression in energy density. With each new fuel source, we have then made incremental improvements in the efficiency of the machines that burn it.

The invention of machines powered by coal began to transform the world in the eighteenth century, and the transformation was consolidated through the efficiency improvements of the nineteenth century. Oil was discovered in the nineteenth century, and it transformed the world through the efficiency gains of the twentieth. Coal and oil are being ripped from the ground in prodigious volumes and are powering the modern world with increasing efficiency. But what did we discover in the twentieth century that might launch us into the twenty-first? The big energy discovery was nuclear power, but we have found it harder to handle than we first expected. Other than that, we don't have much. We have the photovoltaic cell that can convert solar energy into electricity, and we have a selection of other discoveries that make the collection of wind, hydro, geothermal, and biomass energy more efficient, but these are old technologies. They are not transformative in the way that coal and oil were. So what are we left with? It seems that our path into the future is deeply dependent on further efficiency gains to extend the life of old technologies.

59

Planes, Trains, and Automobiles

When the term *efficiency* is mentioned we tend to first think of the fuel efficiency of cars, and it is with the purchase of a more fuel-efficient vehicle that many people make their apparent contribution to the environment. There is a significant rebound effect with car choice, however, as we have seen. People who switch to a more efficient car tend to drive more. This is not a big deal if only one person behaves this way, but it has profound ripple effects when everybody does. More cars hit the roads, more roads are needed, and this paves the way for yet more cars. Many other industries are added to the mix as more tires, more parts, paints, steel, plastics, are needed. One of the main forces spurring the development of the auto industry has been, and always will be, efficiency.

But there is so much more, of course, because the growth of the auto industry does not influence only the industries that are directly linked to car production and maintenance: the modern world has become a veritable car culture, and the automobile has transformed our social landscape. James Howard Kunstler explains this wonderfully in his books and the movie *The End of Suburbia*,[1] in which he gives a compelling description of a culture freed from the shackles of the grimy city by its newly affordable personal transportation, allowing people to head out to a new, fake "countryside" in the boom years after the Second World War. The end of suburbia, as Kunstler describes it, will come when the mobile lifestyle becomes increasingly problematic in an oil-depleted world and the suburbs threaten to become poverty traps.

A gallon of gasoline—a mere jug full—can be made to do an enormous amount of work in an efficient car. Imagine you pour a jug full of gasoline into your sedan and head out to visit Auntie Edith thirty miles away with your family of four. It takes no effort at all, and you are there in half an hour. But oops, you forgot to bring a second jug and the car is now out of gas. Picture the trip back home. You, your partner, and the kids (I hope they are grown kids, not infants) have to push the car those thirty miles. This will give you an idea of the prodigious power embedded in a gallon of gasoline. It's just a jug of liquid that can be held in one hand, and yet it has the power to move a half ton of steel at sixty miles an hour for thirty miles.

An odd thing happened in August 2011. For the first time ever, the US government mandated fuel efficiency standards for trucks.[2] It was a milestone, quite a coup for the Obama administration, and it was lauded by environmental leaders. Oddly enough, the standard was promoted by the truck companies as well.

Big rigs will be required to "reduce fuel consumption and the production of heat-trapping gases"[3] by up to 23 percent, small diesel trucks by 15 percent, and other trucks and vans by around 10 percent. The CAFE[4] standards for cars were also increased in 2011, from 27.5 to 30.2 miles per gallon. This is hailed as major progress for the environment because of all the oil we will save, and as good foreign policy because it will "reduce our reliance on foreign oil."[5] This is the conventional wisdom, but these efficiency standards will do no such thing. The volumes of gasoline and diesel consumed have never been reduced by increasing fuel efficiency standards. The standards have gradually risen over the years, and so has our consumption. CAFE standards were 18 mpg in 1978, 24 mpg in 1984, and they now sit at 30.2 mpg.

And, in any case, do these efficiency standards actually represent all that much progress? The Model T Ford had a fuel efficiency of around 20–25 mpg: not as good as the 2012 Ford Focus, but similar to the 2012 Ford Escape, and much better than the 2012 Ford Expedition. What we've done in the last century is make cars bigger, faster, and more comfortable. Significant efficiency gains have made this possible, but fuel has not been saved.

Airplanes first flew barely a century ago, and they were limited in their scope until jet engines and aviation fuels reached a sufficient level of efficiency. It is only since the 1960s that air travel has become common. Now, of course, it is routine for many people. An extremely high level of efficiency is demanded by air travel in the first place, which is why it took us so long to master it. Before planes could work, they needed to be light, strong, and powerful: a classic efficiency challenge. Planes have become increasingly efficient over the years, but has this reduced fuel consumption? No, of course not. Efficiency has made air travel increasingly affordable. More efficient engines have powered the industry, enabling the production of increasingly fast, large planes, and increasing the demand for more runways and better air traffic control.

Airliners are actually remarkably efficient. If you calculate efficiency in miles per gallon per passenger, even the humongous Airbus A380, the world's biggest airliner, can be as fuel efficient, per passenger, as the Toyota Prius.[6]

The Prius can carry five passengers, but it often carries only the driver. The A380 gets less than a tenth of a mile per gallon, but it can carry more than eight hundred passengers.[7] The A380 is efficient in a number of ways. It uses a high proportion of advanced materials to maintain structural strength with the minimum weight. But, efficient or not, the consumption is prodigious. The A380 has an 84,000-gallon fuel capacity so it can fly across the world.

A case in point is the special "green" and "efficient" development initiatives at Chicago O'Hare, one of the largest airports in the world. The airport boasts a number of green initiatives, such as a small solar array, improved recycling in the terminal, and the judicious use of on-site landscape materials, and these are all commendable, but one thing I find a little dubious is the claim that new runways improve the airport's efficiency. The argument is that the placement of the runways, coupled with better streamlining of tarmac routes, reduces taxi and idle times, saving energy. That may be true, but the real reason for an overhaul at O'Hare is to reduce delays in the country's air traffic system. The new runways increase capacity and provide a better runway to accommodate the Airbus A380. As usual, the efficiency improvement has increased energy consumption.

One of the most significant impacts of efficient cars and planes has been the gradual displacement of railroads, which are energy efficient but less versatile than cars for point-to-point travel and slower than planes over long distances. Train transportation remains a staple for freight, but the demise of the passenger train and our difficulty in reviving it reveals a lot about our new economy. Fuel consumption could be reduced significantly by reducing our reliance on cars and rebuilding train connections, but it would be difficult to revitalize passenger trains while personal transportation and road systems remain so efficient. Such a rebuilding could be of great value in the future, as oil depletion begins to bite, especially given the versatility of light-rail systems to run on electricity as well as diesel, but progress has been slow.

Transportation will be one of our biggest challenges as we pass peak oil because the alternatives for liquid fuels are sorely lacking, but trying to save the world with more efficient planes, trains, and automobiles is a losing proposition. Efficiency will not slow down our consumption of oil. Rather, it will help us to keep scraping as quickly as possible toward the bottom of the barrel, until the barrel can, rather suddenly, no longer be scraped.

How Fridges Changed the World

> Over the past decade, EnergyStar has been a driving force
> behind the more widespread use of such technological
> innovations as efficient fluorescent lighting, power
> management systems for office equipment, and low standby
> energy use. [EnergyStar] products deliver the same or better
> performance as comparable models while using less energy
> and saving money.
>
> —US Environmental Protection Agency[8]

Refrigeration was the first example I used to teach the efficiency trap. I was giving a lecture explaining the need for energy efficiency and alternative energy technologies in an environmental science class, and I decided to take a few minutes to briefly explain the Jevons paradox. I hadn't really thought much about it yet, so when a student asked if it only applied to efficiency in technologies such as the steam engine I had to wing it.[9] "I don't know. Let's pick something at random," I foolishly suggested. "How about energy efficiency in household appliances, let's say, fridges?"

I didn't really expect the discussion to amount to much, but there's much more to fridges than meets the eye. You'll recall the pivotal role of ice-cooled railroad cars in the development of the meatpacking industry. The ability to cool meat catalyzed the emergence of that industry even before the advent of the electric refrigerator, and it ushered in a suite of pervasive socioeconomic changes that spanned the American landscape, both rural and urban. Electric fridges are a much more recent technology, but they have contributed more than you would think to the development of the modern world, and they have done so very rapidly. Fridges are a perfect example of the efficiency trap in all its forms.

First, let's quickly cover the question of direct rebound effects, which are quite obviously present. The very first electric fridges were horribly wasteful little monsters. They were so inefficient that the cooling unit had to be built in a separate box outside the refrigeration unit because of all the heat it generated. And they were tiny things. As refrigerators have become increasingly efficient, they have also grown in size. The average refrigerator of the 1970s could hold less than half as much as the average refrigerator today.

As it turns out, the average refrigerator has always consumed more or less the same amount of electricity. Much less per cubic foot, yes, but about the same overall. The rebound, then, is roughly 100 percent. We return all the energy savings of more efficient refrigerators by going big. And then comes the backfire. Whereas fridges were a new thing only two generations ago, and many families were just acquiring them, the average American home now has more than one monster fridge (and freezers—the logical extension of efficiency refrigeration technologies). So, with far more, far bigger fridges, efficiency has enabled us to increase our household electricity consumption significantly, and this is only the beginning of the surprising power of the humble refrigerator.

Supermarkets and Walmart Supercenters, with their long rows of meat-, milk-, pizza-, and ice-cream-packed freezers, have displaced the local green-grocer, butcher, baker, and fish monger over the last half century. This has been made possible by the efficiency with which large stores can store and sell large volumes of refrigerated and frozen goods, and so the refrigerator has played an important part in the reorganization of food retail. The efficiency of large food retail outlets paved the way for new efficiencies in wholesale, which was also made increasingly efficient with refrigerated trucks. Soon, there was no economic advantage in obtaining food locally. Why grow vegetables in the Midwest when you can grow them more cheaply in California and Florida and truck them across the country? My home state of Indiana, in the heart of one of the world's great breadbaskets, produces vast quantities of feedstock but virtually no food. Refrigeration has contributed not just to the efficiency of households, but to the restructuring of the agricultural landscape.

Refrigeration is a classic example of the efficiency trap. The more efficient our refrigerators become, the more electricity refrigeration consumes and the more its system-level effects grow. Refrigeration uses prodigious amounts of energy, and efficient refrigeration has shaped our kitchens, our neighborhoods, and our landscapes. So in order to avert an energy crisis and reduce emissions of climate-altering greenhouse gases, we should go out and buy an Energy Star refrigerator? I don't think so. And what about an Energy Star heater, washer, dryer, or air conditioner? Well, I don't expect the discussion of those would amount to much . . .

How Many Engineers Does It Take to Change a Lightbulb?[10]

What would shed light on the efficiency trap better than, well, lights? Here is another place that efficiency, we are assured, will save energy. There are various mandates around the world requiring the transition from incandescent lightbulbs to the more efficient compact fluorescents (CFLs). Also on the horizon are the even more efficient LED lights.[11] Compact fluorescents have been successful, and have been adopted by many people, but they have not made quite the splash they were expected to. They tend to be a little slow to warm up, emit a slightly less pleasing light, and cost a fair bit more. Thanks to the energy savings they offer, CFLs are cheaper over their life span, but the saving does not seem to be quite enough to convince everyone to convert. And have you noticed that where the shelves used to be filled mostly with 60 Watt (W) incandescent bulbs they are now mostly filled not with the 60 W equivalent CFLs (15 W) but with the 75 W equivalents (20 W)? That's a 25 percent rebound, right there. Heck, you're saving so much with the CFLs that it's no big deal to get the slightly brighter one, right? CFLs also carry environmental concerns because many contain mercury.

LEDs have much more promise than CFLs. They can operate with as little as a tenth of the energy of CFLs, and they last three times as long.[12] They are much more efficient than current lighting systems. These savings and improvements might well be enough to tip them over the edge into common usage. There are a few technical problems that remain to be cleaned up with LEDs, but these problems should be solvable, and probably in ways that will make them valuable for new applications. LEDs may pave the way for much more easily integrated "smart light" systems and a range of new applications, such as in greenhouses and for sensor-assisted streetlights and the like. There is, as always, a bunch of energy to be saved through this efficiency improvement, but will more efficient lighting save energy, as promised, or is this another efficiency trap?

Roger Fouquet and Peter Pearson performed a detailed survey of lighting technologies over the last seven centuries.[13] They estimated that the efficiency of lighting increased by about 25 percent over the eighteenth century, with the price per lumen roughly halving and the total consumption of light increasing by about ten times. The consumption of candles and whale oil increased, I assume, but the biggest changes were just arriving with the Industrial Revolution. The cost of a lumen fell by a further 75 percent through the nineteenth century, its efficiency increased by fourteen times, and the consumption of lighting increased by more than two hundred times. Things changed much faster in the twentieth century, of course, with oil, natural gas, and nationwide electricity distribution. Through the twentieth century the efficiency with which we generated lighting increased by a thousand times, the price of lighting plummeted, and our consumption rose by more than twenty-five thousand times (with per capita consumption rising by 6,500 times).

There has clearly been an enormous rebound throughout the history of lighting. As lighting has become more efficient, the cost of its delivery has decreased and our use has increased. We consume massive amounts of energy on lighting in the modern world, and our consumption has been driven by efficiency gains. Efficient lighting has had far-reaching impacts on society. Lighting from whale oil lamps and candles was cumbersome and expensive so, in the eighteenth century, we simply went to bed when the sun went down. Now we work through the night in a cheaply, conveniently, and efficiently lit room (or office, factory, or oil rig).

The historical trend line is clear: efficiency has led to consumption, but what will the trend be over the next few decades? Will the efficiency gains of CFLs and LEDs save energy this time, or will they represent another efficiency trap? The tendency, as we have seen with transportation and refrigeration, is for efficiency to either increase consumption or mask declining energy supplies. But perhaps we already have all the lights we can use.

NASA has published a fascinating map that shows how earth looks from space at night.[14] Our planet has a veritable glow. When the sun goes down, the lights come on and you can very quickly figure out who has access to efficient lighting and who does not. The developed world is ablaze with white dots of light tracing the outline of continents and glowing as hotspots around major cities. The Northeast corridor, from Boston through New York to Washington, DC, is a fat, shining worm of light; Tokyo, London, Paris, and Sydney are gleaming pearls. Major thoroughfares are ribbons of light studded with sparks along the length of California, where Interstate Five connects San Francisco and Los Angeles, and through England, where the M6 links London to the center of the universe (Manchester).

But there are distinctly dark places on the planet at night. Some are to be expected, such as the heart of the Sahara Desert, the Tibetan Plateau, or Antarctica, because they are rather empty of people. But there are also huge swaths of darkness where many people live. You might assume, from its virtually black night sky, that Sub-Saharan Africa is devoid of people or, from its subdued lights, that Central Asia is relatively uninhabited. The sky at night shows, very clearly, where the people with access to efficient, affordable lighting live. It's the perfect visual of the efficiency trap. But the sky at night also shows us that efficient, affordable lighting has not yet reached the entire population.

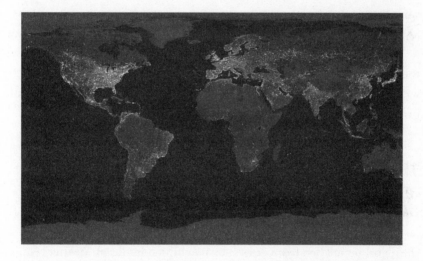

The earth at night. This remarkable image was compiled from multiple images taken from a defense meteorological satellite during 1994 and 1995. The image does not give us a particularly good visualization of where human population is the densest. Rather, it shows us where the most electricity is consumed, which shows us where people have access to the most efficient electricity production and distribution systems. Efficiency gains in lighting systems do not save energy in the long run; they cause more lights to shine. *Courtesy of NASA's Visible Earth project (http://visibleearth.nasa.gov); data courtesy of Marc Imhoff of NASA GSFC and Christopher Elvidge of NOAA NGDC. Image by Craig Mayhew and Robert Simmon, NASA GSFC.*

Imagine that maps of the earth at night were available as a time sequence tracing the history of humankind. The first lights would begin to flicker as the tiny red-point sources of open fires in East Africa about a hundred thousand years ago, and they would spread around the world as fire-wielding people migrated out of Africa. They would reach the coasts of Australia about sixty thousands years ago and then creep across Beringia about fourteen thousand years ago. Concentrations of these dim, red lights would coalesce and brighten on the Nile, Euphrates, and Indus Rivers, then in the Mediterranean and elsewhere, marking the rise of empires. The lights would dim—much more suddenly than they had lit—as these empires fell. About six hundred years ago, the darkness of Europe would begin to break and then the illumination would increase at the dawning of the Industrial Revolution. As coal begins to be drawn from the ground and put to work, the view of earth from space would change much more quickly, and then, in barely a century, the earth would suddenly illuminate into what we see today.

What happens next? Will the night sky shine even brighter, and will the lights come on in parts of the world that they have not yet reached, or will the lights blink out, putting us to bed again, each night, in the dark, beneath a blanket of carbon dioxide?

The Efficient City Trap

One of the grand challenges of the modern world is an exploding population, now in excess of seven billion, and an even more rapid growth of cities. While the global population has increased by two billion in the last quarter century, the rural population has hardly changed at all, and nearly all the population increase of the last few decades, and the increases predicted for the next few decades, have been, and will be, in cities. The rich countries were already strongly urbanized by the middle of the twentieth century and the rest of the world is urbanizing rapidly.

Some countries seem to think, since wealth and urbanization are linked, that urbanization drives economic growth. China seems to have bought into this false cause and effect in a big way. The rate of urban growth in China is overwhelming (while the quality of some of the construction is decidedly underwhelming). I've seen incredible changes in Chinese cities over the last decade. This year, I saw a row of at least two 130-story apartment buildings under construction in an economic development area north of Tianjin.[15] What was amazing was that they were all at exactly the same stage of construction. When finished, I'd guess that's enough apartments to create an instant neighborhood of nearly a quarter million people.

Just one small part of an immense all-at-once apartment block project outside Tianjin, China. *Courtesy of Steve Hallett.*

And it's not just the areas of obvious economic boom that are seeing this growth. In the decidedly back-of-beyond northwestern Chinese city of Xining, my colleague David Umulis was shocked at the city's skyline. He had spent a few months there in 2001 teaching English and had no recollection of any tall apartment buildings whatsoever. There were by 2012 at least five hundred of them. There is no doubt about it. Our population is still growing—and heading to the city.

When you start to look for efficiency traps you will see them everywhere, and it's very disconcerting. Better appliances, better cars, better farming, better planes, better lights; they all result in more consumption. It's obvious, really, once you've had time to think about it, but when you put all these efficiencies together with the increasing consumption they cause, it has a simple name: it's called progress. But, since there is a general clamor among the environmentally aware for efficiency in specific technologies—the car, the furnace, the refrigerator, the lightbulb—it seems logical that there should also be a movement for efficiency in the systems within which these things interact, and there is. There is a whole raft of organizations and movements pushing for the development of more efficient cities. The William J. Clinton Foundation is trying to link business with investment opportunities for efficient buildings, the LEED program of the US Green Building Initiative certifies energy efficient buildings,[16] and there are many other groups investing in efficient building design and construction and efficient transportation networks. There is an Efficient Cities Network[17] and an Efficient Cities Initiative.[18] The theme of the 2010 World Expo in Shanghai, China, was "Better Cities, Better Lives," where myriad technical innovations designed to make cities more efficient were on display. Vast amounts of energy can be saved, it is claimed, by establishing efficient cities.

The most comprehensive analyses of cities are probably those of Geoffrey West and his collaborators, who take a fascinating systems approach. Like me, West believes that there are fundamental rules that can be applied to deepen our understanding of all manner of systems, from organisms to ecosystems, including human systems such as cities. The most accessible introduction to West's findings is on display in a lovely TED lecture, "The Surprising Math of Cities,"[19] in which he demonstrates some of the shared trajectories of biological and urban systems.

West has correlated the size of mammals, as one example, with a range of aspects of their functioning. It turns out that you can predict a lot about a mammal species just by knowing its size.[20] Larger species, such as the giraffe or elephant, require a very large, powerful heart to pump blood but have a slow heart rate. Smaller species, such as rats and mice, tend to have much smaller hearts that beat away like crazy. The metabolic rate of animals tends to be correlated with size, as is gestation period and many other things. West has shown a rather tight relationship between size and many physiological characteristics, which is essentially a relationship between size and physiological efficiency.

What's particularly interesting about the relationship is that while it is very direct, it is not linear. A linear relationship would mean that as size doubles, energy demands would double, too. What we actually see is that as size doubles, energy needs increase by only three-quarters. In other words, biology has an incredible economy of scale. The energy demands of an elephant are orders of magnitude less than the energy demands of the same weight of mice; that is, thousands of mice. It also means, of course, that mammals can get only so big. An elephant twice as big, or a giraffe twice as tall, would be able to move only at the speed of slug.

Cities scale in remarkably similar ways to mammals. Cities are scaled-up towns, which are scaled-up villages, and, just like mammals, as cities double in size, their energy demands increase by only three-quarters. The length of electric lines and roads, or virtually any infrastructure requirement, even the number of gas stations, increases in the same, efficient, nonlinear way as the city gets bigger.

This is the core of the efficient-city argument. As cities house more and more people, they do so with increasing economies of scale, reducing per capita needs. And there is truth in this—indeed, there is mathematical proof. And there is an even more encouraging relationship in cities. The bigger they get, the more investment they attract, and so wealth goes up, this time with the opposite relationship. It's not only that wealth increases with size, but that it increases faster than linear. As a city doubles in size, its wealth production more than doubles.

It would seem that all would be well; that we can have bigger and bigger cities that produce more and more innovation with greater and greater efficiency, but there is a catch, as West explains:

> The catch is that this system is destined to collapse. What we do is, as we grow and we approach the collapse, a major innovation takes place and we start [growing] again . . . and so there is this continual cycle of innovation that is necessary in order to sustain growth and avoid collapse. The catch, however, is that you have to innovate faster and faster and faster. So the image is that we are not only on a treadmill that is going faster and faster, but we have to change the treadmill faster and faster. We have to accelerate on a continual basis. And the question is: can we, as socioeconomic beings, avoid a heart attack?[21]

It is repeatedly stated that efficient cities are the hope of the future. The exact opposite is the truth. The increasing size of cities is inevitable, but cities are anathema to sustainable development. They are huge sinks for resources, and the more efficient they become, the more efficiently they gobble up those resources. With less waste, as always, yes, yes, but with more and more consumption, and the constant growth of cities—especially efficient ones—cannot be maintained forever.

It should be obvious that an efficient city would become a massive consumer. An efficient city would be one that combines all manner of individual efficiency measures in streamlined networks, and the conventional wisdom would see all these efficiencies compounding into a superefficient system. But the efficiency traps also compound. An efficient city is destined to become an enormous black hole into which people, energy, and resources will efficiently flow. An efficient city will work better as long as its demand for resources can be met. But the resource flow will always get bigger and resources must always arrive faster. If the needed resources do not arrive, the city will fail.

The argument against efficient cities is not an argument against better cities. It's not bigger, more efficient cities that we need, but better, more livable ones. A much higher priority should be placed not on city designs that increase efficiency under the current models of transportation and centrally produced electricity, but on those that will continue to support the well-being of people through the tumultuous times that confront us.

There is no particularly great solution to the dilemma of our growing population, nor is there a great solution to the problem of urbanization, but the first step is to recognize that the increasing concentration of people into cities will create huge and intractable societal problems over the next century. In a world of depleting fossil fuels and declining natural resources we may simply be unable to support these overly large resource sinks, and the problem cannot be solved by attempting to reduce consumption through efficiency. Our cities will need to make do with reduced levels of consumption and they will probably need to also generate far more of their own resources. Significant design overhauls will be needed to do this, and we'll take a closer look at some of the options in the last part of the book.

CHAPTER 5

THE ENERGY SUBSTITUTION TRAP

We are desperate to believe in miracles. Technology will save us.
Capitalism is good at technology. So let's just keep the show on
the road and hope for the best.

—Tim Jackson, *Prosperity without Growth:*
Economics for a Finite Planet[1]

The greatest leaps forward have occurred when a society has tapped into a new energy source. The society flourishes as the efficiency with which it harnesses energy improves, and all would seem to be well until it becomes overgrown and overly dependent on that energy source. More efficiency is demanded, but it becomes a trap. The society fails when its energy source can no longer support its demands. A new source of energy is needed, and if it does not come, failure is inevitable. The society pulls out all the efficiency stops in a vain attempt to maintain its growth, but it fails, and not gradually: it collapses.

The modern world is extremely vulnerable. It seems all-powerful, at first glance, but the failed civilizations of the past probably seemed all-powerful, too. It doesn't take much imagination to realize that the twin crises of fossil fuel depletion and global climate change, accelerated by soaring populations and the depletion of soil, water, and biological resources, might represent an existential threat. It takes considerably more contemplation to understand that the oft-promoted efficiency solution is no solution at all, but that's a truth we must accept, too. Efficiency improvements will do nothing to extend the fossil-fuel era or curtail our pervasive environmental pillage. So what other options do we have?

Perhaps new technologies will save us. There are, after all, abundant sources of energy for the taking. The sun sends vast amounts of energy to earth every day that can be harnessed by well-designed houses, photovoltaic cells, solar water heaters, and solar reflectors. The energy of the sun is harnessed by plants. The heat of the sun lifts water into the sky, where it condenses into clouds, then falls as rain, and when it flows down rivers, we can harvest it once more with hydroelectric power plants. The heat of the sun also warms the air, which swirls into vast flows that can be collected with turbines. Geothermal energy is abundant in a few places, and there are certainly untapped sources that could be exploited that would make a considerable difference in those areas.

The biggest known source of non-fossil-fuel energy is nuclear energy, and there is, at least in theory, much more potential for nuclear energy than our current use indicates. There are many complex issues with nuclear energy, however, and I'll get to those at the end of the chapter.

Taken together, the energy available from the sun, the wind, and flowing water exceeds our needs by orders of magnitude. Technology, surely, can cure our fossil-fuel addiction and deliver the energy we need to keep us on the path of progress while cleansing the environment.

At the end of the day, substituting coal, oil, and natural gas with wind, hydro, and solar is not a question of energy availability but of efficiency. Efficiency is something we're pretty good at, but how far can it take us? And where will it take us?

There are two intertwined efficiency trends: the inherent efficiency of the energy source, and the efficiency with which it is harnessed. We tend to focus more upon the second trend. Which car uses gasoline more efficiently? Which furnace uses natural gas more efficiently? The first question, however, is much more fundamental. Which energy source is more efficient: sunlight, wind, water, wood, coal, oil, or natural gas? There are technologies that can harness any one of these sources more or less efficiently, but only within the inherent constraints of the source, and each source is constrained in its own way.

Piles of coal and dung on the roadside north of Hohhot, in Inner Mongolia, China. We burned dung long before we started burning coal, and we'll be burning dung for a long time after the coal is gone. Let's face it: we humans love to burn stuff. Burning whales was a favorite trick of ours for a while. *Courtesy of Steve Hallett.*

Energy Returned on Energy Invested

> Now it will be easily seen that the resources of nature are almost unbounded, but that economy consists in discovering and picking out those infinitesimal portions which best serve our purpose. We disregard the abundant vegetation and live upon the small grain of corn. . . . No possible concentration of windmills, again, would supply the power required in large factories or iron works. . . . Petroleum has, of late years, become the matter of a most extensive trade, and is undoubtedly superior to coal for many purposes, and is capable of replacing it.
>
> —William Stanley Jevons, *The Coal Question*[2]

Comparing energy sources is a little like comparing apples and oranges. Sunlight, the flowing wind or water, a seam of coal, a reservoir of crude oil, or a seam of tar sands; all these have very different characteristics. They have different energy densities, different transportation and storage issues, and offer different opportunities to be converted into other forms of energy and work. Coal, for example, is a solid that can be loaded on to rail cars but cannot flow through pipelines. Natural gas can be piped rather easily, but it's difficult to load on a truck.

Gasoline begins as crude oil in the ground; electricity begins as coal in the ground; photons of light hit a PV cell; wind or water turns a turbine. Fuel ethanol begins as a corn or sugarcane plant in farmland.

Figuring out which energy source is better or worse than another can be confusing and can depend on a number of things, but the most important characteristic of an energy source is our ability to convert it, one way or another, into useful work, and so there is one fundamental question: How much energy must be expended to harness the energy source and put it to work? Howard Odum was one of the first people to struggle with this concept, and he came up with the concept of *emergy*, which measures the amount of real energy delivered. Emergy includes not just the energy source at hand, but also the energy contributed by other environmental systems. Charles Hall, who was a student of Howard Odum's, coined the terms *net energy*, *energy return on investment* (EROI), and *energy returned on energy invested* (EROEI). EROEI is the clearest of these, to my mind, and the concept is extremely important. We too often ask the irrelevant energy question, how much energy is there in a given source? What's much more important is how much of it we can use. There is an immeasurable supply of energy in the universe in the stars,[3] for example, but we can't yet lasso any of it for human consumption. There are massive amounts of energy coming to the earth every day at light speed from our own star, but we're not all that good at getting our hands on that, either.

We have always been surrounded by, and bathed in, vast amounts of energy. We have lived in the burnable forests, on the banks of flowing rivers, on windy plains, and under the warming sun. The problem has never been a lack of energy sources. Rather, the problem has been deficits in our capacity to harness them. Our societies exploded to life when we developed the capacity to harness coal and then oil because we suddenly had huge sources of energy that could be tapped at little cost.

The concept of EROEI gives us the ability to prepare some kind of meaningful comparison of different energy sources, and this analysis has been performed by a number of different researchers whose results I have compiled in table 1. What really matters is not how much energy is out there but whether or not we can afford to put it to work, or, as Henry Groppe put it, "There is no such thing as scarcity and no such thing as surplus. There is only price,"[4] echoing the words of William Stanley Jevons a century earlier, "The whole question of the exhaustion of our mines is a question of the cost of coal. All commerce, in short, is a matter of price."[5]

Table 1. Energy Returned on Energy Invested (EROEI) for selected sources of energy. The ranges represent the findings of different researchers, and also various variables such as, for wind turbines, the quality of the wind at the site at which they are located. Some of the data have significant variability, such as those for natural gas, which combine the findings for both tight and conventional plays.[6]

US Domestic oil, 1930............... >100:1	US imported oil, 2010.............10–18:1
US coal, 1930 ....80:1	US domestic oil, 2010 ....7–12:1
Oil: global average, 1990.........30–40:1	Wind turbines 4–10:1
US coal, 1970 ....30:1	Oil from shales............................ 3–5:1
Firewood 30:1	Solar Photovoltaic.......................2–8:1
Natural Gas 10–80:1	Oil from tar sands.......................2–4:1
US domestic oil, 1970 ....20–40:1	Solar Concentrating
Oil: global average, 2010.........18–20:1	Collector 1–2:1
Hydroelectricity.......................10–30:1	Biofuels 0.7–1.7:1

One of the most striking things is the way the EROEI for oil and coal have fallen over the last century. Although we still have significant reserves, the end of cheap fossil fuels is rapidly approaching. There was a time when we could drop a well down into a huge reservoir and oil would come gushing out. The giant and supergiant wells (of the kind we are not finding any more) consumed only a single barrel's worth of energy to pump a hundred out: an EROEI of 100:1. It now takes a lot more energy to get our hands on the black gold. This should come as no surprise when we consider that we are now producing oils from shales such as the Bakken formation in North Dakota, from tar sands such as the Athabasca formation in Alberta, and from the deep ocean off the coasts of Brazil and West Africa, and in the Gulf of Mexico.

There have been trillions of barrels of oil on the planet throughout the history of humankind. Until we had the technological prowess to extract and refine it economically, however, it was useless. There will remain trillions of barrels of oil on the planet long after the end of the petroleum interval, but it will be too expensive to exploit the oil reserves, so the question is: At what point does it become futile to drill for oil? We don't know the answer to this, but we can be fairly certain that the world's production of oil will be barely a trickle in another century. It is not the end of oil that threatens our civilization. It is the escalating costs of obtaining the oil and the realization that, eventually, we will be too energy poor to produce the oil that remains in the ground.

The recognition of declining EROEI brings home the message of our looming energy crisis much more effectively than estimates of how many billions of barrels of oil we have left or how much sunlight is falling from the skies. The quantity of energy is irrelevant. It is our capacity to harness energy in a way that can power our civilization that counts. The next obvious question, then, is: What EROEI is required to sustain a nation or a civilization (specifically, ours)? If our EROEI continues to decline, at what point do we fall off the net energy cliff? I don't know the answer to this, but an EROEI below 10:1 looks plenty scary.

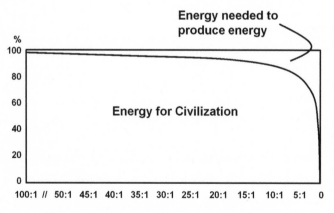

Energy Returned on Energy Invested [EROEI]

The net energy cliff. Civilization collapses not when it runs out of energy but when it can no longer afford to make the cheap energy it needs—and it's not the financial cost of making energy that matters the most, but the energy cost of making energy. As the quality of our energy sources declines, we will eventually fall off the net energy cliff. *Courtesy of Steve Hallett.*

The concept of EROEI seems to have been completely ignored by neoclassical economists. Since there remain trillions of barrels of oil in the ground, they assume that market forces will find a way to produce them as the price increases. The same argument is made for renewable energy: when the market demands renewable energy, it will spur innovation and supply what we need. This assumption is completely, and frighteningly, false.

Increasing price certainly acts as a demand pressure, but declining EROEI quashes it at the same time. You would think that the high price of oil would be spurring investment in oil exploration all over the place, and that the rate of discoveries would be increasing. Despite the hoopla in North Dakota, investment has declined, and the reason is simple. The EROEI for oil *discovery* has fallen through the floor. It has become uneconomical to even look for oil because the crappy, hard-to-produce oil fields that are now being found do not pay back their exploration costs. EROEI is a much more fundamental force than the price at the pump. It is when the EROEI for oil production falls that oil fields are abandoned even though they still have oil in them. It will be when EROEI has fallen globally that the petroleum interval will be over, no matter how much oil is left in the ground.

The Fundamental Weakness of Renewable Energy

When it comes to renewable energy I am both a proponent and a skeptic. I think we need to get much more serious about developing renewable energy, but I have no expectation that it will be nearly enough to preserve our existing way of life. Wind, solar, hydro, and geothermal energy are relatively kind to the environment (not completely benign, but relatively so), but they need to become much more efficient, and they need to be deployed on a much larger scale than seems feasible. The amount of electricity generated from renewable sources has increased significantly in the last decade or so, but it still represents only a small fraction of our total use. In the broadest analysis, as much as 16 percent of the world's energy comes from renewables, but nearly two-thirds of this is wood and most of the remainder is hydroelectricity. Excluding wood and hydro, less than 1 percent of our energy comes from renewables.

Renewables are the fastest growing source of energy, but their rise has been from a very low baseline. What's more, our overall energy consumption, and fossil-fuel consumption, in particular, is also increasing. There is no indication that renewables are replacing fossil fuels.

I think we tend to be seduced by the technological hype coming from our energy specialists. Barely a day goes by without some new claim that another advance, already in development, is about to transform the future of energy. Wind-turbine engineers have doubled the efficiency of turbines, and we now have turbines that can generate electricity in stiff breezes rather than only in near gales.[7] That sounds promising. Solar cell manufacturers are promising high throughput production runs of cheap, machine-printed solar voltaic cells, and our electrical engineers promise us new-generation batteries that will store twice the electricity at half the weight of the best ones on the market today. Maybe they will achieve these goals, and maybe they won't. I hope they will. These advances are extremely important because we need all the environmentally benign energy we can get, but despite all the technical wizardry and hyperbolic claims of imminent energy security, even these advances would not be enough, and the reason is simple: they will never be efficient enough—and perhaps that's a good thing.

Take solar energy, for example. The American economy consumes more energy than is fixed from the sun over the entire landmass of the lower forty-eight states.[8] This is the scale of the energy demands that we have. To fulfill our energy demands entirely by solar energy—although I know that's not the goal—we would need to capture more energy in solar cells than is currently captured by all our forests, prairies, and farmland put together. It's simply not feasible. Indeed, you'd better hope it's not feasible. The countryside would be riddled with solar cells and the environmental impacts would be enormous. Think of the production of heavy metals that would be needed, as just one problem.

The biggest solar farm in the United States is the 380-acre Copper Mountain solar facility in Nevada that has a rated capacity of 48 MW from 775,000 photovoltaic panels. This is a prodigious project, but, put in context, it merely demonstrates that solar power has a problem. It is lauded as a renewable energy breakthrough capable of providing enough electricity to power fourteen thousand average American homes. Nevada's Governor Sandoval announced, "This exemplifies my goal of making Nevada the renewable energy capital of the country. Projects of this magnitude provide hundreds of jobs and invest millions of dollars in our state."[9] Wait a minute. Fourteen thousand homes? Is that it? That's the size of Calais, Maine, or Ketchikan, Alaska . . . or the ancient city of Uruk, Mesopotamia in 3500 BCE.[10] It's just not a game changer. The actual output of solar farms, thanks to the unreliability of various factors such as cloud cover and dust collection on panels, rarely surpasses 20 percent of the rated output, and this solar farm generates less electricity than a small coal-fired plant. And it requires 775,000 technology-dense solar panels on 380 acres in one of the sunniest place in the world to do that.

This idea that the sunniest places in the world are the best sites for solar farms is misleading. One problem is that desert areas, intuitively earmarked for solar arrays, tend to be remote, requiring the generated electricity to be transmitted long distances, and another problem is that deserts tend to be very dusty. A more surprising problem is the fact that photovoltaic cells are heat sensitive and their output consistently declines as temperatures rise. High levels of sunlight and sparse cloud cover are obviously an advantage in terms of light capture, but the sunniest places in the world also tend to be the hottest. My friend Jody Tishmack has a solar system that powers her house in Indiana and she consistently gets the highest output from her system in March. By July and August, when the sun is strong and high, her electricity production falls by 20 percent because it is also hot.

At the end of the day, the best way to generate significant net energy from sunlight is still through plants. It would seem that the mechanism used by plants, which generates electricity only very fleetingly in the chloroplast before quickly converting it into hydrogen fuel (how cool is that?) and then the chemical energy of sugars, is fundamentally superior to the photovoltaic approach that attempts to handle the electricity directly or store it in batteries. This efficiency gap in the generation of electricity from photons of light may also explain why small, domestic solar water heaters are efficient and effective: because they convert sunlight energy directly into heat stored in water, which is also its end use. After decades of high-tech development, the best ways to use sunlight energy are still the ancient ones: photosynthesis and passive.

Solar power has huge weaknesses because it attempts to harness a dispersed form of energy, and no matter how much its efficiency improves, it will always be dogged by this. The answer is not to attempt to emulate centralized power plants; rather, it's to develop solar energy where it has some potential and limit it to applications that are small and dispersed like the energy source itself. A patch of land where fast growing trees can be grown for firewood is a good way to generate net energy locally. Solar water heaters can reduce household energy demands in many settings. A rooftop or an abandoned city lot is the best location for a PV panel because it will generate electricity where it is needed.

Wind turbines suffer from some of the same scale issues that limit solar photovoltaics. They do much better, but they need to be located in areas with reliable, strong winds, and the best sites are often quite far from population centers where the electricity is needed. Wind speed is extremely important, and although turbines have much-improved efficiency in low wind speeds, the amount of energy carried on the wind does not follow a linear relationship. Wind energy is proportional to the cube of the wind velocity; steady, strong winds are needed to make large amounts of energy: indeed, a half gale is the best wind speed. Turbines will still turn in a breeze, but their production drops precipitously at anything below 10 mph, a speed that is not maintained for a long time in most locations. Like solar power, wind power is not capable of replacing fossil fuels using the same centralized system that we currently operate, and that's probably a good thing. What seems like a benign source of energy now, because turbines are few and far between, would, if it were to meet our demands, become a much bigger environmental problem.

The luster of solar and wind energy comes from the nature of the energy sources that power them. Sun and wind are clean, green, and free, and there is no comparison between a fresh summer breeze and a nasty lump of coal. Sun and wind are the epitome of environmental beauty while fossil fuels are the epitome of dirty pollution. But photovoltaic cells are not green. They usually contain cadmium or arsenic, and while sunlight is renewable, photovoltaic cells are not. They have a finite lifetime and have to be disposed of. Electricity storage is essential for most solar applications because of the variable nature of sunlight, and this is usually achieved with batteries. Batteries contain thoroughly nonrenewable and polluting concoctions and have a lifetime of only about five to ten years. Solar energy might be renewable, but solar energy systems are not.

Problems also exist with wind turbines. Have you ever seen one of these things being transported along the highway or being erected? It takes some serious, energy-guzzling machinery to get these things in place. A large turbine costs somewhere in the order of $1 million to $2 million to build and erect, and it takes more than a decade to pay for itself. Wind turbines do have a longer lifetime than solar cells, but they don't last forever. We actually don't really know what the lifetime of modern wind turbines will be because they are, well, modern, but it is probably in the order of twenty to thirty years.[11] Its energy source is fully renewable, but the turbine is not. Modern wind turbines are very large machines, and the blades, with blade-tip speeds in excess of 200 mph in a good wind, generate enormous torque stresses. Regular maintenance is required, which also costs money, and there have been dozens of turbine failures and collapses around the world.

There are a number of environmental concerns with solar and wind power that are commonly brushed under the rug because, despite these failings, it is true that they are much more environmentally friendly than fossil fuels. But their fundamental problem cannot be ignored: the problem of energy density. The energy that solar cells and wind turbines harvest is not a dense-packed bundle of explosive energy, like a rail car of coal, a barrel of oil, or a pipeline of natural gas, but the decidedly dispersed photons of light streaming down from the sun or the scattered molecules of air moving on the wind. No matter how efficient these technologies may become, they will always be faced with the basic problem that lots of energy must be invested in cells or turbines to concentrate the dispersed energy of the sun and wind. They are encumbered with an efficiency dilemma that fossil fuels do not face and the improvements they would need in order to power a civilization like ours are unlikely to come.

Denial Is a River in Egypt[12]

The most productive form of renewable energy in the world today is hydroelectricity, representing the vast majority of the world's portfolio of renewable energy and about 16 percent of the world's total electricity generation. Most of this comes from large, river-based hydroelectric dams such as the Three Gorges project on China's Yangtze River, which, when completed in 2009, became the most productive power plant in the world; the famous Hoover Dam on the Colorado River, the first huge dam of the modern era, built in the 1930s; and the prodigious Aswan High Dam on the Nile in southern Egypt, which transformed that country's economy by nearly doubling its electricity-generating capacity. Brazil generates 85 percent of its electricity from hydro. Norway generates effectively all its electricity from hydro.[13] Even the United States, clearly a fossil-fuel-powered country, generates a solid 8 percent of its electricity from thousands of hydroelectric dams scattered across the nation.[14]

While most of our renewable energy options have failed to materialize in any substantial way, hydroelectricity has performed well, and it has done so for decades. The critical advantage of hydroelectricity is, as always, the availability of energy in a compact form. Water does not have the inherent characteristic of high energy density, but water moving downhill does, and hydroelectric dams are basically devices that concentrate the power of flowing water. A river is dammed, trapping water and raising its level in a reservoir, concentrating the gradient along which the river flows from a gradual incline to a sudden, precipitous drop. As the water flows through the dam its potential energy becomes kinetic energy, and turbines are turned. The larger the volume of water trapped and the greater its height above the turbines, the greater the amount of electricity that can be generated.

Scale presents significant engineering challenges: the recently completed Three Gorges Dam is made from a veritable mountain of concrete (enough to build a sidewalk that would wrap around the globe six times, apparently) and took decades to complete. A large hydro project requires a huge capital investment. The city of Yichang, at the dam site, was a small town until it was transformed into a booming city by the massive influx of workers. Scale aside, though, hydroelectric projects are pretty simple. Water flows. Turbines turn. The biggest problem is finding good sites for dams, and this, quite simply, is why hydro will not add much capacity in the future. The world is long past peak dam because most of the good sites have already been dammed.

As far as efficiency traps are concerned, hydro is relatively free from the problem of escalating consumption. In this case, you do want to consume as much of the river's kinetic energy as possible, so you want the dam to be as efficient as possible. There are no problems here, but hydro projects cause significant efficiency traps at the ecosystem level.

When Egypt completed the prodigious Aswan High Dam across the Nile in 1970, it doubled its electricity production and gained a number of other significant benefits. The dam was able to control the water level of the Nile, preventing serious droughts and floods. More sophisticated irrigation systems became possible around the river once its level was stabilized. Lake Nasser was formed at the head of the dam and new fisheries developed there. Additional water was able to flow into water-scarce Sudan. The Aswan High Dam, in short, was a major boon for Egypt. Lots of emissions-free energy was generated, agricultural production increased, death and property damage from floods was prevented, and a fishery was created. The project seemed like a wonderful win-win proposition, but now that the dam is nearly fifty years old, the system-level problems are becoming increasingly clear.

Things are not as they used to be in Lake Nasser. The crocodile population has exploded, for one thing, which makes life a little more interesting. The lake is also gradually silting up and increasing in salinity, which makes the water less suitable for irrigation. On the downstream side of the dam, where the erratic flows of the river have been evened out, the calmer waters are now more suitable for *Schistosoma*-carrying freshwater snails, and the incidence of the debilitating disease bilharzia has increased. Long-term changes are also becoming apparent in the huge Nile Delta, which is shrinking due to reduced siltation and increasing wave damage from the Mediterranean Sea. Fisheries in the Mediterranean Sea have collapsed because the outflow of the Nile has slowed.

The impacts of the Aswan High Dam on agriculture have not yet fully been felt because Egyptian farmers have been using cheap fertilizers to secure and boost crop yields, but the Nile system is one that has been shaped for many thousands of years by a huge river that would flood periodically and bring replenishing nutrients from the highlands. This was recognized by the ancient Greek historian Herodotus in the fifth century BCE. Those nutrients no longer come from nature; they must be replaced by technology.

Is the Hummer Greener Than the Prius?

To all you Prius owners out there: I'm sorry, but this is going to hurt. Yes, you are the good guys because you're trying to help, and you are likely helping in other ways, but I'm afraid there's more to the world's most famous energy-efficient car than meets the eye.

To all you Hummer drivers out there: don't be smug. You bought the stupid thing for all the wrong reasons. It turns out that you haven't done as much environmental damage as you thought you had, but you are probably still an ass.[15]

To all you Toyota executives out there: I'm not picking on Toyota. This applies to all hybrids. You still get props from me for being one of the best car companies in the world. I've owned four Toyotas, and they were all great (well, I bought the first one for $500 back in the day, so let's say it was as great as you could hope).

It's shocking to be even contemplating the idea that a Hummer might be as green as a Prius because the Prius gets better than 45 mpg and the Hummer gets a paltry 13 mpg.[16] The simple math, then, shows that, after 120,000 miles of driving, the Prius has burned 6,500 fewer gallons of gasoline than the Hummer. That's a lot of gas. And it has emitted a lot less carbon dioxide: roughly 60 fewer tons. How on earth could the two cars be even remotely comparable?

Well, if all we consider is the emissions of the vehicles over 120,000 miles of driving, there is no comparison at all, but if we consider the entire life cycle of the vehicle from manufacture to disposal, things look quite different. Most of what I'm writing here comes from CNW Research's "Dust to Dust Automotive Energy Report,"[17] which gives a very detailed life-cycle analysis of cars. There are numerous critics of this report, however, so consider digging deeper if you want fuller details.

The complexity of the Hummer-versus-Prius controversy starts to emerge when we consider a number of other characteristics. The manufacturing costs of the Prius are unusually high for a car its size, and it is energy intensive to build. The car is assembled from parts manufactured all over the world and shipped to assembly plants. The manufacturing costs are also high because the Prius uses more advanced materials than most cars in order to keep its weight down. Another factor in the "Dust to Dust" report, although this has been contentious, is the lifetime of the Prius. Hummers are likely to go nearly one hundred thousand miles farther than Priuses, on average, before being scrapped. This is significant because it means more Priuses would be manufactured and scrapped per mile driven. Add to this the costs of repairs and maintenance in the not-so-standard Prius, with its advanced parts, and you begin to see that the Prius is not quite the green machine it seems to be.

The biggest issue with the Prius is its advanced nickel-halide battery, which is ironic because it is the hybrid gas-electric powertrain that provides the high fuel economy and is therefore the biggest attraction of the Prius. Thousands of tons of nickel are mined and smelted near Sudbury, Ontario, which presents a significant pollution source in that area. The nickel is then shipped to Wales for refining and then to China for further processing before being shipped to the battery plant in Japan. That's a lot of shipping, requiring lots of energy and emitting a lot of carbon dioxide. The completed batteries are then shipped on to assembly plants. The disposal costs of the Prius are also higher than those of the Hummer, again largely due to the disposal of the batteries.

When we consider all these factors, the Prius-versus-Hummer calculation is much closer than you'd think. I'm pretty confident that the Prius still wins, but the paragon of green driving is really not that much better for the environment than the monster of the roads.

It's fun to be argumentative by comparing the Prius with the Hummer, but the Hummer is something of a special case. Relatively few people drive these monsters. Much more instructive is to compare the Prius with something more similar. The Toyota Corolla is roughly the same size (a slightly smaller interior) but costs about $5,000 less that the Prius. It doesn't have nearly the same green image, but it's actually much, much greener. The Corolla is cheaper to make, cheaper to maintain, runs farther, and is cheaper to dispose of. It has less of a green image because it gets about ten fewer miles per gallon than the Prius, but over its lifetime it consumes less than half the energy of the

Prius. The Corolla contains no fancy battery and therefore does not cause the additional problems of nickel mining, smelting, refining, and disposal. The Corolla is more than twice as efficient as the Prius with fewer environmental side effects. The Prius steals the green image that the Corolla deserves.

The Prius-Corolla comparison shines light on our awkward relationship with efficiency. It's hard to save energy in one part of a system without it busting out in another. Hybrid cars, at the end of the day, are still gasoline cars. The battery is charged when the car brakes, which is, indeed, rather brilliant, and an onboard computer system switches from gas to electric to maximize efficiency. They do use gasoline with extraordinary efficiency, but they require extra components, manufactured at extra expense, as well as energy to achieve this. So, while hybrids have fantastic gasoline efficiency, they do not have particularly great energy efficiency overall.

The next step up in efficient cars is the plug-in hybrid and the electric vehicle (EV). According to a project conducted by ECOtality, "Around one million gallons of petrol and 8,700 metric tons of carbon dioxide emissions have been saved by EVs in the US in the last three years"[18] Let's take a quick look.

The concept of the EV is much more revolutionary than that of the hybrid. The EV is an electric car, so it uses no gasoline or diesel whatsoever. It runs on electricity, which is a big advantage because the conversion efficiency to motive force is higher for electric engines than for gasoline engines. But electricity is not a source of energy: it's only a carrier of energy, and electricity needs to be generated first. So here's your first problem: from what source of energy is the electricity generated?

If I were to buy an EV here in Indiana, where most of our electricity is generated by coal-fired power plants, driving an EV would basically mean buying a coal-fired car. Sure, I'd save the gallons of gasoline promised, but the carbon dioxide emissions? Not so much. The emissions reductions would be significant if the electricity were from a renewable source, such as hydroelectricity, but not from coal. The environmental potential of EVs, then, is heavily reliant on the economic potential of renewable energy.

There are a number of other problems with EVs. Like the hybrids, they rely on advanced batteries, in this case, lithium-based rather than nickel-based, but there is still the problem of lithium mining and smelting—and other heavy metals are used in batteries, such as cobalt, nickel, and copper, with the same problems. There are also the same issues of disposal. The batteries can be recycled, in theory, although this process is not effectively in place, and they can be down-cycled to other uses. Again: not in place. Last are the issues of value. A Chevrolet Volt will set you back twice as much as much more capable equivalents, and it will have a much shorter range.

The capability of EVs is a huge problem that remains to be solved. The Nissan Leaf claims a range of one hundred miles, but in tests carried out by *Car and Driver* magazine, the range was only fifty-eight miles.[19] We also know that the deep-discharge batteries in EVs, while they last a long time without failing, do tend to lose capacity quite quickly, reducing the range of the car significantly in just a few years. Users also found the recharge time of at least twelve hours annoying. If you take it out for supper in the evening, you might not be able to use it to drive to work in the morning. People found themselves leaving the AC, heater, and other electrical gadgets turned off to save charge, and what happens if you get stuck in traffic for a couple hours? *Car and Driver* concluded that the Nissan Leaf "might make a good second car," but the EV reveals the continuing problems of the efficiency trap.

Purdue University's solar racing team, a student organization, has come a long way since its first vehicle, the Boilermaker Special, built in 1993. By 2008 their Pulsar vehicle had a fuel efficiency equivalent of nearly 5,000 mpg. That's amazing, but it has a top speed of only 26 mph, weighs in at a mere 158 pounds, and can't carry much more than the smallest licensed student drivers on campus. Their latest car, Celerias, has upright seats, looks a bit more like a car, can cruise comfortably at a little over 30 mph, and has a range of over 400 miles in the right conditions.

Transportation presents the most difficult challenge in the future because our current system is deeply dependent on liquid fuels from oil. Now that we have had a good few decades investing in alternatives we can summarize our future options fairly easily. Hybrid vehicles are not a game changer. They are actually less efficient than gasoline vehicles and are on the way out soon. Electric vehicles have developed well, and will continue to improve, but they will always be very limited in their capabilities. They are much more efficient for light loads and short or slow trips (we use them on golf courses all the time), but they are impossibly inefficient beyond unconquerable limits of cost, weight, range, and speed. If electric vehicles are to be the future of transportation, our lives will be lived much more locally and long distance transportation will be much more expensive.

This superefficient, clean, green, prototype electric smart car is plugged into the grid in Shanghai, China. So it's a coal-powered car. *Courtesy of Steve Hallett.*

The latest supercool solar car designed and built by students at Purdue University. You can go for a long, superefficient drive around the great state of Indiana at a comfortable 35 mph. What you can't do is take your mates and camping gear with you. *Courtesy of Steve Hallett.*

Why Biofuels Are for Biofools

> There is surely nothing quite so useless as doing with great
> efficiency what should not be done at all.
> —Peter Drucker, "What Executives Should Remember"[20]

Biofuels are viewed by many people as both a potentially important fuel of the future and an environmentally friendly source of energy. They do represent a potentially useful fuel, but not a source of energy, and they represent a significant source of environmental hazard. First, there is a distinction between biofuels and biomass, both of which are sometimes lumped under the term bioenergy. Biomass is just a fancy term for plant material that can be collected for the production of biofuels or simply burned. Burning biomass, whether wood or grass, has been a source of energy throughout the history of humankind, and it is not complicated to understand its pros and cons.

Wood was the dominant source of energy for every preindustrial civilization.[21] Civilizations flourished as they found more and more efficient ways of putting forests to the axe, and a number of them fell as a direct consequence of deforestation.[22] The equation is simple. Wood must be consumed no faster than forests can regrow, and the consequences of the efficiency trap are clear. Increasing the efficiency of wood harvesting can represent an existential threat. The hunt for firewood is a major cause of desertification at the southern margins of the Sahara Desert, and of landslides in Central and South America, where trees are required to hold soil in place to prevent erosion by wind or water. It makes a lot of sense to burn grass and wood for fuel, but only if it can be done sustainably. The Finns get a good proportion of their household heating from sustainable forestry. Compacted grass or wood-chip pellets are good options for heating in many places.

Biofuels are a completely different concept. Here, the goal is to convert biomass into a liquid fuel, and the first problem is obvious. It adds extra processing steps. Rather than simply collecting the biomass and burning it to release energy, the biomass is instead converted into liquid form and the liquid burned. Energy is lost in the process, and so, while it is relatively simple to generate a net energy gain from biomass, it is more difficult to generate a net energy gain from biofuels.

Why we want to convert biomass into liquid form is, of course, obvious. While all sources of energy can be used to generate either heat or electricity, only oil is really effective in vehicles, especially large vehicles, ships, and planes (though natural gas shows some potential).[23] Gasoline, diesel, and aviation fuel, all distilled from oil, may not be essential to keep the lights on, but they are essential to keep the wheels turning. Similar liquid fuels can be made from biomass, but is energy actually gained from biofuels when we consider the amount of energy used to make them in the first place? What is their EROEI?

Biofuels can be made from two broad categories of biomass: high-value plant tissues, such as seeds, or low-value stems and leaves. The high-value plant tissues are harder to grow and harvest while the low-value tissues are harder to process.

Nearly all the biofuel produced to date comes from high-value plant tissues. In the United States, about a quarter of the Midwest corn harvest now goes to the production of bioethanol and a smaller proportion of the soybean harvest goes to the production of biodiesel. A number of oilseed crops are used for the production of biodiesel in a number of countries. The world's biggest producer of biofuels is Brazil, producing bioethanol from sugarcane.

The volume of biofuel produced globally is now fairly significant, representing nearly 2 percent of liquid fuels consumed globally,[24] but only Brazil seems to be actually generating any energy this way. The rest of the world is either producing barely a trickle of fuel or else is producing it at a loss. Corn bioethanol would seem, at first blush, to represent an energy bonanza. Twelve billion gallons of bioethanol were produced in the United States last year, representing nearly 5 percent of the gasoline market, but was energy actually made, or was it just converted from one form to another?

Corn is not just a wild plant that can simply be harvested from nature. It needs to be planted and fertilized; weeds and pests need to be managed; and then the corn needs to be harvested and distributed to refineries. And corn is a greedy crop. When you add up the diesel used in tractors, combines, and trucks, along with the natural gas used in fertilizer, the energy equation begins to look highly dubious. The harvested corn must then be processed into ethanol and the ethanol itself must be distributed.

There have been a number of analyses of the corn bioethanol energy cycle. The numbers vary a fair bit depending on the breadth of the life-cycle analysis performed, but few analyses demonstrate real potential for bioethanol. One of the most outspoken critics of biofuels has been Cornell University's David Pimentel, who has performed detailed analyses and concludes that it is almost always a net energy loser.[25] Even if Pimentel is overly negative, as some critics claim,[26] the energy balance is so tight that ethanol is a virtual nonstarter as a source of energy. Even if its EROEI is above one, it is only by a whisker. We could not power the modern world by burning wood, let alone corn. Why would we think we could power it by squeezing fuel from corn kernels?

The other claim of biofuels proponents is that they reduce carbon dioxide emissions. The argument here is that the carbon dioxide emitted by the burning of the biofuel is offset by the carbon dioxide fixed by the growing plant in the first place. As with the net energy argument, it makes sense at first blush but falls apart when you consider the full life cycle of the process, accounting for the carbon dioxide released from all the fossil fuels involved in the process and the nitrous oxide emitted from the fertilized land. The biofuels industry is a significant greenhouse-gas emitter, and the more widespread and efficient the industry becomes, the more it will emit.

The energetics of corn is fairly disastrous, but sugarcane is a little better. It grows more rapidly, requires less management, and stores a high concentration of easily fermented sugar that can be relatively easily collected. Because the energetics of harvesting ethanol from sugarcane is better than that for corn, Brazil, at least, is actually making energy as well as just liquid fuels. It does come at a significant cost, however. The expansion of the cane lands pushes the expansion of soybean production (used mostly for animal feed but also for biodiesel) into the cerrado, a unique South American grassland ecosystem that has been systematically converted to cattle rangeland. The cattle, also in demand, then push into the rain forest. Sugarcane ethanol is not grown on land cleared of rain forest, but it does not take much imagination to understand that bioethanol is a significant indirect cause of the rapid and disastrous deforestation of one of the world's most important biomes.

The easy biofuels—those generated from high-value plant tissues—are relatively easy to produce, but they are dogged by two problems that cannot be ignored. They generate no energy and they threaten widespread environmental damage. They will also tighten the grip of industrial agriculture in North America on unnecessary feedstock corn and soybean production. At the end of the day, all we are doing is converting fossil fuels into biofuels at the expense of the environment. These biofuels are only for biofools, and the scientific community has pretty much accepted this now.

The other route to biofuels is to use low-value plant tissues. The logic here is that we can avoid using quality land that would be better reserved for food production by instead using plants that require little management. Any chunk of old cellulose will do, such as fallen tree limbs from city streets or swaths of prairie that will burn at some time anyway. First of all, let's not ignore the obvious efficiency trap. These are the benign materials that could be used, but, should the technology actually ever work, you can bet your life people would want to get their hands on better materials than these. They'd be back in our farmland. I don't think we need to worry about it too much, though, because converting plants into gasoline just won't work. Again, we could never power our modern world by burning wood, let alone wood scraps and prairie grasses. Why would we think we could power it by squeezing liquid fuel from them?

There are two methods proposed to do this. The first is to hydrolyze and digest the cellulose and then ferment it to ethanol, and the second is to simply squish it into liquid by various combinations of chemical and thermal means. Neither method makes any thermodynamic sense, and, again, even if the EROEI sneaks above parity, it will be by only a whisker.

In a world without oil we are going to have to get a lot more of our energy from plants, but it will come in the same two ways it always has: by burning plants and by eating them. The small amount of liquid fuels that we may generate will make sense only for a few specialist tasks in just a few locations.

Efficient Oil and Gas Production: The Big Fracking Shale Mess

We sit at the peak of the petroleum interval, this great glitch in the history of civilization. The big fear is the imminent decline of oil and natural gas supplies, but, apparently, there is an oil and natural gas boom going on. The global picture is grim, and peak oil is upon us (peak gas to follow in a few years), but there are still some significant oil and natural gas booms in a few places. One such place is Williston, North Dakota, which sits above the Bakken oil-shale formation. Locations in Pennsylvania, Ohio, and Texas also sit above massive gas-shale formations.

Oil and gas shales are examples of what the petroleum industry calls "tight plays." While conventional oil and gas reservoirs are located in loose source rocks through which the oil or gas can flow freely to be collected at the well, shale oil and gas must be retrieved from reservoirs in which the oil or gas is found in dense source rocks. You might picture conventional oil as water that drips freely from a wet sponge and a tight play as water held in a block of clay. Give the sponge a gentle squeeze, and out comes the water. Squeeze the clay, and you just get sticky hands. It is much harder to produce oil from tight plays than from conventional ones, since it requires much more energy to do so. As a result, the petroleum industry has struggled to exploit oil and gas shales economically.

The breakthrough in exploiting shale formations was made by Texas petroleum engineer George Mitchell,[27] whose innovation was to drill horizontally through the shale formation and inject a gel-like or foamlike fluid called a *proppant* into the well to cause cracks.[28] The process is known as hydraulic fracturing, or fracking. The opening of multiple fractures through the shale loosens the formation, allowing gas or oil to find its way to the horizontal well through which it can be removed.

Mitchell's innovative technology, coupled with higher oil and gas prices, has made fracking economically viable in many places. The petroleum industry has declared this to represent the discovery of vast volumes of recoverable oil and natural gas. An example is the now oft-cited statement made by President Obama and others, that the United States now has another hundred years' supply of natural gas *at current rates of consumption*.[29] The American Natural Gas Alliance airs those really convincing TV ads with the really convincing-sounding lady who walks convincingly over a large American floor map to convince us that natural gas is now the "domestic, abundant, and clean energy" that can power our lives. The Bakken shales have been hailed as the biggest oil discovery in the last three decades.[30]

The volumes of oil and natural gas that will actually be recovered from shale sources, as well as the rate and duration of recovery, however, are sources of significant debate. There will certainly be a fair amount of oil and gas coming from these sources, but how much, exactly? The estimates of recoverability from the Bakken shales range from an encouraging 50 percent to a decidedly discouraging 1 percent. Likewise, the estimates of recoverability from the gas shales range considerably and could be either very significant or decidedly "meh." Seldom mentioned is the fact that most of the gas comes from a very few "sweet spots" in the shale formations, and it's unclear how the rest of the formations will perform.[31] I think the truth is probably closer to the lower range of estimates rather than the higher range, but that's not really the point I want to analyze here. Of much more interest is the very fact that we are now finding it necessary to scrape oil and gas out of this nasty Nutella jar. The store has clearly sold all the good ones, and that's a very big deal. Gas production in the United States is not declining, and that's largely thanks to fracking, but it's not increasing, either. Like the Red Queen in *Alice in Wonderland*,[32] we're running faster and faster just to stay in the same place. We will take quite a tumble when we fall off this particular treadmill.

Tight oil and gas plays are an environmental nightmare. The Athabasca tar sands in Northern Alberta are extracted by surface mining and the use of vast volumes of natural gas–generated steam. They produce poor quality oil,[33] and for each gallon of oil they separate from the sands they generate two barrels of oily, sandy goo that gets dumped into massive sludge ponds. Unconventional oils are much more polluting than conventional oils.

There has been a rash of problems and complaints about the production of shale gas by fracking.[34] The trouble began with benzene contamination in a water well in Wyoming. Foul odors were detected in domestic water, and skin rashes from showering were reported in Texas. In Ohio, a couple's house blew up when gas from their water well filled their basement. A woman in Colorado blames her health problems on the chemicals used for fracking.[35] The list goes on and on, despite the claims from the industry that fracking is safe.

We already release natural gas to the atmosphere from conventional gas production—actually a shocking amount. Estimates are that more than 5 percent of our natural gas is lost from leaky pipes and about 3 percent is lost during production from conventional fields. Fracking may more than double these losses.[36] This is a serious problem, since natural gas is a much more potent greenhouse gas than carbon dioxide—and it's a terrible irony, since, while natural gas is a much cleaner fuel than oil, its production by fracking might be much, much worse. Nearly a third of the greenhouse gas impacts of the United States may come from natural gas.[37]

So producing oil and natural gas from these tight plays is extremely costly. Much more advanced technologies must be used in the process; much more energy must be consumed; the economic margins are much lower; and the environmental consequences are worse. Why, one might ask, are we doing this? The answer is that we are approaching peak oil, the easy, cheap oil and gas are gone, and we are beginning to scrape the bottom of the barrel.

We can see the problem at many other oil and gas production sites. The Deepwater Horizon tragedy in the Gulf of Mexico occurred because an oil well blew up, but also because the wellhead was at the bottom of the ocean. Why are we taking risks like this by drilling in such difficult places? Because the easy, cheap oil is gone. Why are we pushing to drill in the Arctic National Wildlife Refuge (ANWR) and on the continental shelf? Same reason. Why are we designing oil rigs that can withstand damage from icebergs and freezing seas? Because we're looking to drill in the Arctic Ocean.

Increasing efficiency in the petroleum industry is giving us access to more oil and natural gas, but the scales are beginning to tip. The EROEI of oil has gone into sharp decline as we approach the production peak. We are producing more oil than ever, to be sure, but the rot is beginning to set in. The second half of our global oil and gas endowment, as we pass peak oil, is going to be much more costly, to both our pocketbooks and the environment.

Clean Coal and Other Oxymorons

Pick your favorite: military intelligence; postal service; congressional ethics; civil war; friendly fire; fuzzy logic; sustainable development; jumbo shrimp; train schedule; wilderness management; corporate culture; rap music; clean coal. The idea of clean coal is too silly to take seriously so let's cut to the chase.

Coal is pulverized and blasted into a furnace that heats water to generate steam that turns turbines to make electricity. It's not fancy, but it is cheap. In fact, there's not much to be said in favor of coal power except that it's cheap. Well, you could say it's clean, I suppose, but that would be a lie.

There are two pollution issues with coal-fired power plants. One is the suite of nasties that they release, such as particulates, mercury, and sulfur dioxide. The other is carbon dioxide. Significant improvements have been made over the years regarding the nasties, in particular sulfur dioxide, the cause of acid rain. Little progress has been made with carbon dioxide, which is particularly discouraging because new technologies are available that could enable power stations to operate with significantly reduced carbon dioxide emissions. The most obvious of these technologies is the integrated gasification combined cycle (IGCC) plant that first converts coal to syngas and then burns the syngas to drive the turbines. IGCC plants can reduce sulfur emissions significantly, more easily capture the sulfur dioxide for conversion into downstream chemicals, and reduce other pollutants. It is also much easier to capture carbon dioxide from IGCC plants for later sequestration.

But there are two catches. First, the IGCC plants do not actually dispose of carbon dioxide, they are simply "capture ready." The sequestering of the carbon is a separate problem. In theory, the carbon could be shoved deep into mines or to the floor of the ocean, but this hasn't happened yet and, in my estimation, because of the huge cost, it never will. A few pilot plants have been built to capture carbon dioxide to grow algae for biofuels, but these have not yet demonstrated a high level of effectiveness and they would be very difficult to scale up (and in any case, the carbon is still burned eventually). Being capture ready is not being clean. It's a bit like having a bathtub in the house but no water.

The second problem is the up-front cost of the IGCC plants themselves, which run 20 percent or so more than regular pulverized coal plants. Only a few have been built, with government support, and they have not fared particularly well. The Wabash River Power Station near Terre Haute, Indiana, has been shut down periodically and has repeatedly failed to comply with clean-water codes because of pollutants discharged into the Wabash River. After two decades of touting this technology, we have exactly three plants in the United States, one of which has been abandoned. The other two have struggled.

Cost is a huge issue for coal because it lives and dies on being cheap. As soon as the costs rise, investors would rather invest elsewhere because, while cheap coal makes financial sense, everybody knows clean coal is an oxymoron.

It's actually a real shame, and a rather tantalizing problem. The technology is almost there to create energy from coal and capture the carbon dioxide, and this would be extremely valuable. We still have significant coal reserves in the United States that will become significant volumes of carbon dioxide. My expectation, however, is that virtually none of the capture will ever actually happen. It will never be perceived by politicians, the industry, or consumers to be as valuable as affordable electricity, especially when the energy crunch swings into full force. Coal is cheap, for now, but it will never be clean.

A Nuclear-Hydrogen Future?

> If you ask me, it'd be little short of disastrous for us to discover a source of clean, cheap, abundant energy because of what we would do with it.
> —Armory Lovins, interview in *Mother Earth News*[38]

There seems to be little hope that we can maintain economic growth and our high-paced lifestyle. Fossils fuels will soon be going into decline, and their massive net energy return and versatility cannot be matched by renewables. Renewables collect low-density, dispersed sources of energy and are difficult to scale up. Each one has its place, especially in niche situations and at small scales, but renewables cannot keep the juggernaut of civilization churning forward. And efficiency is no solution. Efficiency will do relatively little to increase their capacity, and it will magnify the problems that they cause. We are left with only one option to consider: nuclear power. If this won't work, we're in trouble. (Spoiler alert: it won't, and we are.)

Nuclear energy has always seemed to have the potential to generate enormous amounts of electricity, but it has not lived up to its promise. It's as if the industry never really grew up. Technologies are supposed to emerge slowly, prove themselves, and then make a few important breakthroughs that launch them into the marketplace. The first nuclear plants were already pretty effective, and it seemed reasonable to expect that nuclear power would reshape the world as the efficiency improvements inevitably flowed. Maybe it has simply not yet happened. There are certainly many proponents of nuclear power who see an extremely bright future. But maybe the potential is simply not there. Splitting or fusing atoms clearly releases huge volumes of energy that eclipse all other forms, but we don't seem to have been able to control the energy in an economically viable way. Why is that?

Meanwhile, we are afraid of nuclear power, and with some justification. Just as people were beginning to get over the disasters at Three Mile Island and Chernobyl, and many people, environmentalists among them, were becoming increasingly supportive of nuclear power, we had the disaster at Fukushima-Daiichi. We are also plagued by the problem of the disposal of nuclear waste and, perhaps even more importantly, the perceived problem of disposal.[39]

Nuclear power has both huge advantages and huge disadvantages, and so it is difficult to predict its future. It has the greatest apparent energy potential, from a pure physics standpoint at least, and yet it faces the biggest obstacles. The World Nuclear Association (WNA), in its "nuclear century outlook,"[40] makes it clear that nuclear energy is the only viable source of energy to maintain global economic growth. It also makes a case for the environmental benefits of nuclear power, since widespread use of nuclear energy would result in a decarbonizing of our energy production. The decarbonization argument is a strong one, and the delivery of nuclear energy through electricity and hydrogen fuel would, indeed, remove carbon from the energy-generation equation. The WNA also claims that achieving these goals is feasible. We have been told of this enormous potential many times, however, and it has yet to materialize.

Nuclear has be the world's slowest-growing energy source over the last few decades.[41] *Forbes* magazine declared that nuclear power plants do not make economic sense,[42] and after the Fukushima-Daiichi disaster, the *Economist* fronted a special issue with the title "Nuclear Power: The Dream That Failed."[43] Despite the hype of new nuclear power plants being commissioned all over the place, the truth is that even more are being decommissioned. While China is adding a further twenty-six plants to its existing sixteen, the biggest investment in the world, fully one-third of the nuclear power plants in the United States may have to close early. Little known fact: twenty-eight plants have already closed in the United States and ninety have closed worldwide. The nuclear industry has been decidedly stagnant for some time. So which is it? Is nuclear energy a potential savior or an industry that will continue to limp along?

The way things stand at the moment, nuclear power does not look at all promising. The electricity generated by nuclear plants is probably the most expensive of all and the investments are simply not coming. Investors are certainly made nervous by the risks posed by massive failures, but they seem to be even more nervous as a result of poor performance. Things will be very different soon, however, as the price of energy continues to climb. Although nuclear energy is clearly not competitive at the moment, will it become more competitive in a world of higher prices?

The answer is yes, but that's only part of the story. The real problem with nuclear energy, and the real measuring stick, as with all energy sources, is EROEI. It seems that splitting or fusing atoms should give enormous net energy yields, but it does not. It does give enormous yields, to be sure, but it also consumes vast amounts of energy and, therefore, results in very low net yields. This, of course, is why it costs so much. The EROEI of nuclear energy will only change if it can be generated more efficiently. It will not change just because the cost of energy can support it. I expect to see more investments in nuclear power in the coming decades, but I do not expect them to be all that large and I do not expect them to change our fate much at all.

Another of our great hopes for a sustainable energy future is hydrogen. Hydrogen is not a source of energy—there are no pools of hydrogen lying around to be dug up—so before it can be used as a fuel, hydrogen must first be made. It is a carrier of energy, like electricity.

The big dreams for hydrogen are in transportation. It is becoming clear that trains, planes, and automobiles are going to really struggle as oil goes into decline because of a lack of liquid fuels and the limitations of electric vehicles. Perhaps hydrogen holds the key. It is actually much more energy dense than petroleum-derived fuels and it has the added benefit of being carbon-free. Wood and coal contain lots of carbon, oil contains a little less, natural gas (CH_4) contains very little, and hydrogen (H_2), of course, contains none. Hydrogen, on paper, should be the future of fuel. It is better for the environment and packs more punch.

There are, however, enormous technical problems with hydrogen, and I think these will make it too expensive to develop in the way some would like. Hydrogen is the smallest of the atoms and it is also highly explosive and corrosive. It is extremely difficult to contain and is highly prone to leaking. It cannot be moved through existing pipelines, such as those used for natural gas, and I think the costs of deploying it will remain prohibitive. Expensive handling will lower the effective EROEI of any energy source deploying hydrogen, and the energy sources we are moving to in the future are already struggling to break even. Hydrogen can be generated from wind, solar, and hydro, but I don't think much of it will be generated at all.

Because hydrogen has to be made, it is at a massive disadvantage to gasoline, which is easily and cheaply made from oil, and this is where nuclear power is supposed to come in. Hydrogen can be made from water by any energy source that can generate electricity, so sufficient nuclear capacity would give us both electricity to keep the lights on and hydrogen to keep the wheels turning.

I don't think it's going to happen. There are too many problems with nuclear for it to generate the capacity that would be needed, and there are too many technical problems with hydrogen for it to become a fuel rivaling our liquid fuels derived from oil.

But what if I'm wrong?[44]

Let's even go a step further and say we develop a nuclear-fusion capability. That would give us almost unlimited power, which would seem to be a great thing but would actually be an unmitigated disaster. Why are we so desperate to find the energy we need to keep this civilization alive? Success in solving our energy crisis would merely enable us to trash the planet even more and make our collapse even bigger. It's time to stop grasping at straws.

An Efficiency Bridge to the Future?

We spent two and a half centuries climbing the mountain of civilization to these dizzying heights until we found ourselves in the rarefied air but lost in the clouds. Which way will we go from here? Some think we can keep climbing while others are convinced that this is the peak. Those who try to climb higher will simply find they cannot, and as the storm clouds gather, the peak will soon be a very dangerous place. The only logical path is to climb carefully down off the peak. But, wait: Is that another peak that we can see through the clouds? Perhaps we don't need to climb down at all. Perhaps we can build a bridge across from this peak to the next.

This idea of a bridge to the future seems to be the new goal of many environmentalists. We spent a long time trying to prevent ourselves from getting trapped in this position, repeatedly saying, "If we don't fix this in the next decade . . ." We figured that alternative energy technologies would come along, that carbon exchanges would solve the climate puzzle, perhaps even that public opinion would suddenly tip and everyone would stop consuming. But we didn't fix it in all the various decades we had. The very last gasp seems to be to do an extreme makeover on our efficiency technologies while buying some time by doubling down with the last of our fossil fuels so that we can make them last long enough to build that bridge.

There will be no bridge, and even if there could be, the expanse across to the next, theoretical peak would keep getting wider. The more we push, the more we consume, pushing our fading dreams farther into the distance. I'm sorry, but time's up. It's time to start looking for a path down into the greener valley below.

CHAPTER 6
THE EFFICIENT FOOD TRAP

Man can and must prevent the tragedy of famine in the future
instead of merely trying with pious regret to salvage the human
wreckage of the famine, as he has often done in the past.

—Norman Borlaug[1]

The production of food and the generation of energy are the two most fundamental activities of humans and their societies. Calories from food, for humans and livestock, were the dominant form of energy that supported many societies, so energy and food were once the same thing. Agriculture, then, is history's classic primary industry.[2] But agriculture is a primary industry no longer in the developed world. To be considered as such, it would need to generate energy, but modern agriculture consumes energy, and lots of it. We are no longer converting only the unlimited energy of the sun into calories, but also the decidedly finite energy of oil and natural gas. We are eating fossil fuels.[3]

Good Intentions Pave the Way to Hell

"The industrial food system is always looking for greater
efficiency, but each new step in efficiency leads to problems. The
industry approach, when it has a systematic problem . . . is not
to go back and see what is wrong with the system but to come
up with hi-tech fixes that allow the system to survive.

—Michael Pollan, interview in the movie *Food, Inc.*[4]

105

I should probably be more careful about what I say in this chapter because I am a professor at one of America's great land-grant universities,[5] whose mission, for nearly a century and a half, has been to promote agricultural research and teaching. I am discouraged by much of what has happened to American agriculture over the last half century and I'm worried about where it's headed. My views represent something of a minority opinion among my peers. Most agricultural scientists consider their profession to have performed wonders, and their main source of pride is the enormous increases made in the yields of crops and livestock. There is certainly reason to be proud and impressed. Yields of many crops have more than doubled.

It's an interesting pickle in which we find ourselves. Great progress has been made and yet the treadmill never seems to slow down. A hundred small steps, each of which seemed positive in the short term, each building upon the last, have led us to extremely productive agricultural systems but also down a cul-de-sac that would have been hard to foresee. Despite the advances made in modern agriculture, we find ourselves in the embarrassing situation of needing to support American farmers with subsidies so that they can stay in business, and we have developed an agricultural system that is making us sick. Meanwhile, the number of people starving in the world has not decreased: it has increased. What good have we really done when our abundance has led to an epidemic of obesity in rich countries while a billion people still starve in poor countries? We have raised yields and fed more people, but the treadmill continues to turn, and faster. And it seems that we are headed down a track that calls for more of the same.

We need to be more honest about recognizing where our massively increased yields really came from. Agricultural science has certainly played a role, and part of the explanation for the increased yields has been improved crop and livestock genetics and technologies such as pesticides and improved machinery. Accounting for much more of the explanation, however, is the abundance of cheap oil and natural gas that we have enjoyed. Modern agriculture runs on diesel fuel made from oil and cheap nitrogen fertilizers made from natural gas. If our ability to remain on the yield-boosting treadmill is reliant on our ability to access more and more oil and gas, what happens when they go into decline?

We have also drained our landscapes of water to irrigate our crops. Lakes, rivers, and underground aquifers are depleting in many of the world's breadbaskets. We have also mined a good proportion of our phosphorus supplies, another essential nutrient applied to crops in large volumes. Phosphorus rocks are found in only a few places. The United States and China have been the two big producers (and consumers—China will not allow exports), but their supplies are dwindling. The Saudi Arabia of phosphorus, oddly enough, is Morocco, and its resources are being exported with ever-increasing speed. David Vaccari predicts a precipitous fall in phosphorus supplies in the near future, at the same time that nitrogen and water supplies are failing.[6]

It's all well and good to demand higher yields, but how are we going to achieve them with declining soil quality and water, phosphorus, and fossil fuel shortages?

Why the Green Revolution Failed and the Next One Will, Too

> We summarize the situation by saying: "There is a shortage of food." Why don't we say, "There is a longage of people"?
> —Garrett Hardin, "Lifeboat Ethics"[7]

The drive for continued yield increases in rich countries is repeatedly linked to the need for more food in poor countries. Since there are people starving in the world we are told that we should produce more food anywhere that we can. This view has a long history, and it has fared particularly well in the era of free markets. Countries that cannot produce enough food should, we are told, simply buy it from those that can. But while there is an enormous international trade in agricultural products, most food is still grown at home, especially in poor countries, and the fact that a billion people overeat while a billion starve should be evidence enough that the free-market approach to food distribution is hopelessly flawed. The abundance of the rich simply does not make it to the tables of the poor. There is a huge distribution problem, but even when food is distributed it creates new problems.

The idea of free markets is rather silly in a world whose wealth is as unevenly distributed as ours, but agricultural markets don't even pretend to

be free. Rich countries pump so much support into their agricultural industries that they can sell grains below cost. As a result, the distribution of food
from rich countries to poor has the added problem of undermining the agricultural markets in the poor countries. All countries, or at least regions, need
to be as self-sufficient for food as possible.

Norman Borlaug, the "father of the Green Revolution," recognized this.
Borlaug was a wheat breeder working in Mexico when his innovative research
yielded semi-dwarf varieties of wheat that responded extremely well to irrigation and fertilizer and doubled yields almost overnight. His results impressed
the Rockefeller Foundation, which funded an expansion of breeding into Asia.
At the newly formed International Rice Research Institute in the Philippines,
researchers developed IR8 dwarf rice that doubled or tripled yields of that
crop. A massive production increase occurred in India, and that country, which
had been facing imminent famine,[8] suddenly had a relative abundance. The
Green Revolution was a striking success; it produced the food that was so desperately needed, fed millions that would have starved, and made possible the
later ascent of some of our emerging economies, particularly India's. Borlaug
was awarded the Nobel Peace Prize in 1970 for his work, and he invested back
into agriculture by founding the World Food Prize in 1986.[9]

But the Green Revolution didn't take us off the treadmill. It enabled
us to run faster so that we could keep up, but the world's population is still
growing. The major breakthroughs of the Green Revolution are hailed as
breakthroughs in the science of plant breeding, but the new, high-yielding
varieties were not so much better plants as they were plants that were better
at responding to inputs—and those inputs are running out.

We often prefer to ignore it, but we know that the backbone of agriculture is fertility and water. You can raise wheat in the desert with masses of fertilizer and water from deep fossil aquifers, as the Saudis have done for a decade
or so, but once the water supply dries up (as it is, rapidly, in Saudi Arabia), the
desert will return. The Green Revolution never reached places where irrigation water could not be found, especially Africa.

The prodigious use of irrigation has depleted freshwater sources in many
regions. The rapid disappearance of the Aral Sea under the pressure of cotton
irrigation along central Asia's Syr and Amu Rivers bears witness to the volumes

of water that can be withdrawn by irrigation. Lakes, rivers, and especially underground freshwater aquifers are in depletion in all agricultural regions. Meanwhile, the machinery that supported massive yield increases will also soon need a replacement for diesel fuel. Cheap synthetic fertilizers will soon be a thing of the past. The results will be catastrophic, and there will be no miracle cure this time. Depletion of soils and aquifers is not a quick-fix problem.

The Green Revolution brought huge increases in food production and prevented the widespread famine in India that had been predicted in the 1960s. But did it prevent famine or merely delay it? The population of India has doubled since then, the number of internal refugees such as those living in this community camping on a dry riverbed near Chandigarh is on the rise, and famine looms again. *Courtesy of Steve Hallett.*

How the USDA and Monsanto Killed the Family Farm

> The whole system is made possible by government subsidies to a few large crops like corn. It's a form of socialism that's making us sick.
>
> —Robert Kenner, interview in *San Francisco Weekly*[10]

Sadhu, a migrant worker from Bihar, manages this irrigation channel on a farm outside Hoshiarpur in India's breadbasket, Punjab. It's a very simple system that draws water from underground with a simple electric pump. But hundreds of thousands of simple systems like this are draining Punjab's aquifers and threatening its long-term potential to produce food. *Courtesy of Steve Hallett.*

OK, so before we get too carried away with the blame game, the USDA intended not to kill the family farm, but to support it, and Monsanto and the other agribusinesses have never attempted to kill the family farm either: they are just out to make money. Over the history of modern agriculture, however, rural communities have died, the landscape has turned into massive, monotonous monocultures, and farmers have lost control of their businesses. The last century has seen enormous growth in agriculture, but growth in what? Growth in yields and growth in the amount of money changing hands, and growth in all businesses related to farming . . . except farming itself. Agriculture has grown into a multibillion-dollar industry in which farmers are still farmers, and very few of them are rich. How did this happen?

I had a fascinating conversation with a group of Chinese farmers a few years ago in Sichuan province and they asked me about American agriculture. I told them about the prodigious size of American farms, the huge combines, tractors, and planters, and the cheap fertilizers that American farmers use. One farmer exclaimed that American farmers must be very wealthy. No, I explained, they are not wealthy. Most American farmers need to hold a second job to have a decent level of income to support a family. Well then, continued the farmer, the rain must be very bad. No, I explained, the rainfall is usually more than adequate.[11] Very confusing, he replied, and how many acres did you say these farmers cultivate? Usually about a thousand or so, I responded. Well then the soil must be terribly poor, suggested another of the farmers. No, the soils are probably the best in the world: deep and fertile. Then what is the problem, asked one of the farmers? I support my family very well on two acres. Just imagine what I could do with a thousand!

This particular conversation is a long one, and I've had it a few times now. It takes some time to explain to a Chinese (or West African, Honduran, or Indian) farmer that, on the richest soils in the world, in the richest country in the world, with a perfectly suitable climate and plenty of rainfall, a midwestern farmer requires a thousand acres to make a living. Somewhere in the conversation comes the suggestion that "they must be stupid," but it's not that at all: they are trapped. Some clarity comes to the conversation when you explain that a lot of money is actually being made, but not by the farmer. Ahhh, the government is taking the money! (This is an idea to which they can relate.) Er, no, it's not that either: the government gives money to the farmers. . . .

How *did* this happen?

Earl Butz, a former dean of agriculture at Purdue University, was the secretary of agriculture during the Nixon administration.[12] It was an interesting time in agriculture, and a time when food prices were high enough to generate some political heat. When Butz took office, the old, New Deal–era farm bill was subsidizing farmers to limit corn production in order to maintain prices. He quickly did away with it and instituted subsidies for production. This was also the era when new agricultural technologies were hitting the market and agribusinesses were gaining market share. The famous refrains from Earl Butz were that farmers should "plant fencerow to fencerow" and "get big or get out." Agriculture was mechanizing, modernizing, and corporatizing, and it would lead to a better, more productive system. It actually made a bunch of sense, or at least must have at the time. The trend that was set in motion, however, has led to an unhealthy and dangerously vulnerable modern food system.

Subsidies were instituted only for the commodity crops—corn, soybean, and wheat—and they were paid by the bushel. Since that time, farmers have been motivated to produce higher yields of these crops above all other considerations. This does not seem like an insidious problem, and it certainly would not have seemed so at first, but these subsidies have increasingly warped the value of commodity crops. They have empowered the agribusinesses, whose lifeblood is products that increase yields, and since farmers have clamored for higher yields, they have also clamored for yield-boosting products. The real price of agricultural commodities, meanwhile, has gradually declined, and so, while the businesses have been raking in the cash, farmers have been growing more crops with higher yields on more land for less profit. They have been buying each other out and consolidating into huge ventures, but still they don't get rich. This is a classic efficiency trap, and American farmers have been caught fast.

One of the more visible problems with modern agriculture, and one that has been making lots of headlines recently, is the fact that our food supply is making us sick. The press has scrambled to cover the recent repeated outbreaks of diseases such as *E. coli* in hamburger meat and *Salmonella* in cantaloupes. These outbreaks are going to continue to happen, I'm afraid, because it's impossible to make our food supply completely clean. The way we go about solving the problem, though, is disastrous, and it's what Michael Pollan describes in the quote at the top of the section. Each new outbreak is seen as a new problem to fix, and so we go about fixing it so that we can continue to keep producing food the way we do. What we don't acknowledge is that each problem is really an indication that the entire system is flawed, and each new fix—each new attempt to increase the efficiency with which we control it— digs us deeper. We find ourselves in the ridiculous position of having a food industry so complex, cumbersome, and bound in red tape that only the huge producers can stay in business, which is a shame because the hugeness of the enterprise is a big part of the problem.

Robert Paarlberg, a standard-bearer for high-yield conventional agriculture and agricultural biotechnology, and the author of *Food Politics: What Everyone Needs to Know*, spoke at Purdue University in April 2012, delivering a seminar called "The Culture War over Food and Farming: Who Is Winning?"[13] I'm not sure who the sides are supposed to be, exactly, why this is considered a war, or who is winning, but I was really struck by the way he answered an audience question at the end of the seminar. He was asked what the agricultural industry should do about the public perception that the nation's food supply is unsafe. He said that the agricultural industry should separate itself from the food industry. Agriculture produces commodities that are perfectly safe, he said, and should not be blamed if the food industry turns it into junk food. What? The idea that a farmer should want to distance herself from the food she produces strikes me as a sign that there's a pretty huge problem, especially when, at farmer's markets around the country, organic and sustainable growers are reaching out to their customers and explaining, eye to eye, exactly where and how they grew the food that they are selling, even inviting customers out to the farm to see for themselves.

America is in the throes of a diabetes epidemic, and it is caused by the food we eat. We produce masses of cheap calories and process them into masses of fast food that we advertise aggressively. It's cheaper to fill up on fast food at McDonald's than it is to make a healthy, home-cooked meal. There's something weird about that. And since there are significant parts of the country where it's difficult to get your hands on healthy food and awfully easy to get your hands on cheap fast food, we have a problem. Sure, we can blame people for getting fat, but we can also blame our agricultural and food industries for making it easier to eat badly than to eat well.

Many people seem to blame farmers for the strange shape of American agriculture, but farmers are trapped more than anyone. Many farmers object to any kind of criticism of American agriculture, seeing it as an attack either on their way of life or their livelihood. (It reminds me of the way West Virginians or Kentuckians object to criticism of the coal industry, failing to see that, in the bigger picture, the coal industry has given them black lung, crappy jobs, and a degraded environment.) But most farmers understand what is going on, and many would get off the treadmill if they could. It's just really hard to change direction when you have large capital outlays to keep a big business going, guaranteed markets and subsidies for the one product that you know how to produce, and lots of support in producing it.

Please tell me why it is so essential to increase corn yields. Is it so we can make this corn mountain beside the already-full silos outside Lafayette, Indiana, even bigger? *Courtesy of Steve Hallett.*

So this must be an oil refinery, right, or maybe a chemical plant? Actually it's a food-processing plant operated by Tate and Lyle in Lafayette, Indiana. Corn comes in on the railroad lines and high fructose corn syrup goes out. Just looking at this photo makes me thirsty for a soda. *Courtesy of Steve Hallett.*

The companies have a frightening amount of control over farmers. If you want to raise chickens commercially it's really hard to avoid doing business with either Tyson or Perdue. But the control that Tyson has over its farmers is amazing—and scary. Tyson owns the chicks, sets strict rules on what buildings they can be raised in, decides what feed will be used, and basically dictates the price that will be paid for the chickens when they are grown. Monsanto owns the powerful Roundup Ready technology and charges farmers for the seed (and they must use Monsanto's seed) and the herbicide that goes with it (and they must use Monsanto's herbicide even though there are dozens of products virtually identical to Roundup, since the active ingredient went off patent years ago). Monsanto has a special task force that seeks out and prosecutes farmers who save seed, insisting that they buy new seed from Monsanto every year. Red Gold, an Indiana tomato-processing company, is hardly in the Monsanto control-freak league, but it provides its own tomato varieties to farmers and gives strict guidelines on how they are to be grown before Red Gold buys them back to make paste.

This stuff, I've got to say, annoys the crap out of me. Farmers tend to be proud, independent folk who work hard and live on the land, and the companies so often treat them like serfs to a modern-day capitalist lord.

A student who took my sustainable agriculture class at Purdue recently expressed some of this frustration, telling the story of her family's thousand-acre farm in Indiana.[14] The family is deep in debt and, with land prices extremely high right now, they could sell the place for close to ten million dollars. Why don't they? They are Hoosier farmers; they are proud and hard working, and they want to pass the land down to the next generation. A thousand acres is a lot of land, but this particular student has two older brothers, so she plans to leave home and open a flower shop. "There just isn't enough land to support the three of us." What a bizarre, ironic, and depressing state of affairs.

Another grower close to us wanted to pass part of his farm to his daughter and son-in-law but, again, they didn't have enough land. All they could pass down was five hundred acres, and that's not enough land to make a living growing corn and soybeans, as most farmers do around here. The son-in-law's solution was to start raising tomatoes.

It seems impossible to break out of the system. It is very risky to attempt to jump from the treadmill, but it can be done, and there are more and more groundbreaking farmers doing things differently. Adam Moody produces grass-fed meat on 250 acres right here in central Indiana. Neal Moseley makes a living growing vegetables on twenty acres and in repurposed hog barns. Kevin Cooley produces a massive amount of organic vegetables on just ten acres. Ben Hartman and Rachel Hershberger make a living farming just five acres. We'll meet some of these characters in chapter 12.

Most of these problems with modern agriculture are the natural consequences of unfortunate policies in a capitalist economy. That the system drifted in this direction was not the purpose of any particular agency or corporation. The system made sense to most people at the time, as did each incremental efficiency improvement that followed. Few could have seen where it would all lead—but we must have the wisdom to see it now. Over the last decade or so, corporate meddling in the system has, indeed, started to take a stranglehold. While the drift to exploitation may have begun accidentally, it has become decidedly purposeful. Corporations aim to dominate American agriculture, and some of the ways they do it are decidedly scary.

Corporations have weaseled their way into our public institutions and government in an effort to drive policy to increase their control. One particularly blatant conflict of interest was the hiring of the president of South Dakota State University onto the board of directors of Monsanto. The leader of South Dakota's main public institution, charged with the task of leading that state's research, teaching, and outreach in agriculture, is paid, according to Food and Water Watch, $390,000 a year by Monsanto, a corporation with its own interests in South Dakotan agriculture.

Food and Water Watch went on to detail many ways in which large corporations have increasingly influenced agricultural research, and they are particularly critical of the land-grant universities for allowing corporations to distort their mission.[15] Public funding for agricultural research has declined over the last decade or so to be taken up by corporations. This is not entirely a bad thing, but at what point do public institutions become the research arms of corporations? As Food and Water Watch puts it, "Private sector funding not only corrupts the research mission of land-grant universities, but also distorts the science that is supposed to help farmers improve their practices and livelihoods. Industry-sponsored research effectively converts land-grant universities into corporate contractors, diverting their research capacity away from projects that serve the public good." Interaction among public and private researchers is to be encouraged, but we must be careful to ensure that the public research is enhanced by the interaction, not commandeered by it.

So where do we go from here? We have to face the fact that American agriculture has been captured by corporations and that the agenda of those corporations is not to make the world a better place. We also have to realize that their "more yields, more yields" strategy will line their pockets without leading to a stable, healthy world without famine or malnutrition. We also need to realize that this high-input, high-yield agricultural system that we have drifted into, and become mired in, is depleting our natural resources. It is difficult to get off this treadmill, but we must because it is vulnerable to failure as oil and natural gas production go into decline. Agriculture is caught in a vicious efficiency trap. We strive for efficiency, but of the wrong kind. We convert oil, natural gas, and water into cheap, unhealthy food and corporate profits with breathtaking efficiency. We need to stop that. We need to efficiently convert sunlight into healthy food, and vibrant farms and farming communities.

How Efficiency Spoiled the Organic Movement

The organic-food movement was an early rebellion, of sorts, against the big-ag machine. Its emphasis was on pesticides, the dangers of which were brought to the attention of the world by Rachel Carson's *Silent Spring*,[16] but, in the beginning at least, it was a much broader disestablishmentarian movement.[17] It called on farmers to eschew synthetic fertilizers and pesticides, and it had a deep focus on low-input, sustainable agriculture. It was a throwback movement in many ways, advocating the maintenance of soil quality and fertility through composting, manure, and well-designed crop rotations and cover cropping.

In truth, all agriculture was organic until the advent of synthetic fertilizers and inorganic pesticides. Albert Howard promoted organic agriculture as early as the end of the nineteenth century in Britain; Rudolf Steiner's (sometimes decidedly odd) biodynamic movement began with similar ideals in the 1920s; and J. I. Rodale founded his still-famous Rodale Institute in 1940, but the organic movement had its first real takeoff during the social movement years of the 1960s. It then took off again more recently to become a powerful industry that has captured a significant place in the food market-place. You can now go into most grocery stores and find not only a good selection of high-quality, affordable organic produce, but also a huge selection of processed organic goods, and not just granola bars.

But something has gone wrong with organic farming, and it has become a victim of its own success. It's not altogether a bad thing, to be sure, and it depends on what is important to you. The huge success of the industry has certainly made much more pesticide-free food available, and if that's what you see as the most important thing, you'd have to consider the industry a huge success. But it has come at a cost, and many of the other goals of the organic movement, as it was originally conceived, have frequently been discarded.

Organic agriculture is governed by a strict set of rules, and farmers are formally certified.[18] The rules and the certification process give the organic industry great strength because consumers can trust that what they are getting is free of synthetic pesticides and was grown without the use of synthetic fertilizers. But the rules are also its big weakness. As long as growers stay within the letter of the rules, their produce can be certified as organic. They do not, however, need to stay within the spirit of the rules, nor do wholesalers or retailers. As the organic industry has gathered momentum it has become a lucrative business and has sought efficiency at every step. Synthetic pesticides are not allowed, but a long list of natural or inorganic pesticides are. Likewise, synthetic fertilizers are not allowed, but fertility can still be trucked to the farm.

One of the most efficient solutions to the production of food without pesticides has been to move organic production to locations where cheap labor can be used for hand weeding to replace herbicides. Consequently, you will find that quite a few of your organic veggies come from overseas. That's fine, of course, and they are labeled as such, but shipping food across the world, while it fulfills the rule of excluding herbicide use, neither conserves energy nor develops local agriculture. Another solution to the ever-present problem of weeds is to cultivate more. Again: a pesticide-free solution, but not one that is good for the soil. Another limitation to organic farming is plant disease, which, in the absence of fungicides, can cause huge losses. The solution to this problem has been to locate organic production in areas such as California's Central Valley (which also has access to cheap immigrant labor), where humidity is extremely low and disease pressure is much lower. But veggies require a lot of water, and since rainfall is low, the water must come from massive amounts of irrigation. The produce is organic, but it's not necessarily grown sustainably.

The biggest surprise with the organic industry, however, is its corporate structure. Although the origins of the organic movement, and the image most people have of it, is of small, independent (somewhat hippy) growers, the organic industry is actually highly consolidated and dominated by a few huge corporations. The small organic growers that you might come across at farmers' markets are a very small piece of the organic pie.

There is an urgent need to make farming more sustainable. We need to protect and build our soils, conserve water, ensure that the food we produce is healthy, and revitalize family farms and farming communities. Organic farming took a huge step in the right direction, but it has been subverted into a second form of big ag, and the base problem is an efficiency trap. In terms of productivity and profitability, especially in competition with larger enterprises, it is inefficient to grow food using truly sustainable practices. The land simply cannot be worked as intensively. Farmers are stuck in a marketplace that demands high levels of efficiency, and they can be tempted to compromise the long-term sustainability of the land. Sustainability has been squeezed out of the organic movement by efficiency.

The Dismal Theorem

> The Dismal Theorem: "If the only ultimate check on the growth of population is misery, then the population will grow until it is miserable enough to stop its growth."
> —Kenneth Boulding, *Collected Papers*, vol. 2[19]

Kenneth Boulding's dismal theorem is a well-named masterpiece of bleakness. Population is a very simple function of births minus deaths. If you want population to decline you must either reduce births or increase deaths and there are no direct solutions that are particularly attractive. Much more morally pleasing, in the short term at least, is to raise the bar and allow population to grow. But Boulding was not done with his Dismal Theorem; he expanded it into the even more aptly named Utterly Dismal Theorem: "Any technical improvement can only relieve misery for a while, for as long as misery is the only check on population, the improvement will enable the population to grow, and soon will enable more people to live in misery than before. The final result of improvements, therefore, is to increase the equilibrium population, which is to increase the sum total of human misery."[20]

Right, so the Utterly Dismal Theorem is pretty depressing. There is no technological solution to overpopulation: it can be delayed but not solved. This is the basic efficiency trap created by the productivity ethic in agriculture, and it's the reason that no green revolution can ever win a final victory. A population problem can find no solution in food production because raising food availability to the next level will simply raise population to the next level, too—and that is exactly what has happened over the last half century. Food production boomed through the years of the Green Revolution but the global population more than doubled as well, and the most dramatic population growth has been in the countries that were its biggest beneficiaries. We simply cannot produce our way out of a population problem. The problem always catches up again.

So what can we do? Well, thankfully, Boulding has a third theorem to direct us. His third theorem, with more than just a little irony, is called the Moderately Cheerful Form of the Dismal Theorem: "If something else, other than misery and starvation, can be found which will keep a prosperous population in check, the population does not have to grow until it is miserable and starves, and can be stably prosperous."[21]

It's actually not all that much help because it doesn't tell us what that "something else" might be.

We'll get back to some "something elses" in the last part of the book, but the critical message here is that we cannot produce our way out of a population crisis. This supposed reason for increasing yields, then, can finally be discarded. The efficiency trap in agriculture has damaged our land base, depleted water resources, and consolidated farms into a set of industrial processes.

Meanwhile, agriculture has become deeply reliant on fossil fuels. Wherever it has done this in the name of feeding the world it has done it in vain.

There is nothing wrong with trying to produce as much food as possible from the land. That will always be one of the goals of all farmers, but it should not be the first goal. The first goal should be to maintain the land and its resources for future generations.

PART 3

THINKING IN SYSTEMS

CHAPTER 7
LEARNING FROM NATURE

The ecological thinker is haunted by the consequences of time.
—Garrett Hardin, *Living within Limits:*
Economics and Population Taboos[1]

S ociety has adopted efficiency as a key mechanism for improving the environment by accepting the conventional wisdom that efficiency is equivalent to conservation. It is not equivalent at all, and it tends to increase consumption rather than reduce it. We should be deeply suspicious of efficiency claims, and this is the first efficiency trap we must learn to detect. But the consequences of our constant drive for efficiency are much more profound than this. Efficiency is a veritable engine of growth and progress. It not only fails to check consumption, it also serves a central role in growth and development. This could be viewed as a second efficiency trap: one that is much broader and systemic, causing an efficiency gain to reach out of its own domain into the wider milieu.

These efficiency traps are troubling and, since they run counter to the conventional wisdom that would have us save the world through efficiency improvements, they are a nightmare for the environmental movement. But there are yet more traps. Efficiency does not only increase the throughput of a system, it can also influence its functioning and shape the dynamics of the system itself. In order to reveal these more subtle traps, we will need to delve more deeply into some systems thinking.[2]

The types of systems that concern us in *The Efficiency Trap* are primarily human systems that shape our societies and economies. They include relatively simple, localized systems, such as business cycles or local transportation systems, and much larger systems, such as economic systems and governmental models. In order to pave the best possible path into the future, we need to ask tough questions about those systems and try to understand them fully. Ideally we will be able to understand systems well enough to make robust predictions about how they will perform through time so that we can design them better. All this will also have to be done on the backdrop of a warming world riddled with environmental problems and a looming energy crisis that will, unfortunately, make things harder still.

We hear cries for increased efficiency in all our complex systems. The mantra of efficiency is chanted again and again as a means of controlling systems and making them stable and productive. To keep local bus routes operating they must run more efficiently. To maintain our way of life we must generate economic growth more efficiently. To survive the perils of global climate change and extend the life of our fossil fuels we must use energy more efficiently. To make governments effective, we must make them collect and disburse taxes more efficiently. Efficiency is seen as a cure-all, not only for saving energy and resources, but also for protecting vital social systems. But what role does efficiency really play in complex systems? Does efficiency actually help to sustain them, and is efficiency really a stabilizing force?

To understand the true role of efficiency in complex systems we need to digress for a while and delve into biological and ecological systems. These are systems that have somehow managed to sustain themselves over huge expanses of time and yet maintain an inordinate level of stability. How on earth do they do it?

Why Efficiency Is No Substitute for Sex

> There is grandeur in this view of life . . . from so simple a beginning endless forms most beautiful and most wonderful have been, and are being, evolved.
> —Charles Darwin, *On the Origin of Species*[3]

The Efficiency Trap concerns itself with the sustainability predicament in which the world finds itself today. Its central thesis is that efficiency is not the remedy for a sustainable future that it seems to be. Efficiency is the impetus for consumption and growth and is anathema to sustainability. Efficiency is the laboratory in which the machines of industry are devised; it's the workshop in which the tools of consumption are sharpened. But the perils of efficiency are not restricted to the artificial world of human economies. They can be seen in the natural world as well. Efficiency, unchecked, compels anything, whether a technology, a factory, a civilization, or a species of animal, virus, or plant, to consume with ever-increasing hunger until it brings about its own destruction.

Living things are the most incredible machines, honed for efficiency by millions of years of evolution. We apply our most advanced technologies in an attempt to emulate them but, more often than not, our attempts are clumsy and inelegant. Airplanes and helicopters employ the same basic principles of flight as birds, bats, and insects, but they fall so pathetically short.

The delicate hummingbird hovers with breathtaking precision to dip its proboscis into a nectar-bearing flower and then, in an instant, wheels and darts away. The albatross glides inches over the ocean for hours on a transcontinental voyage with barely a flap of its wings, riding the whisper of updraft that rises from the water. The frigate bird stretches its wings wide to circle and climb on thermals and then, spotting a fish in the ocean below, folds them back and swoops in a sudden plunge. The bat maneuvers silently through the forest in the dead of night, catching insects in flight.

Our advanced flying machines can certainly go higher and faster than birds, bats, and insects, but they lack subtlety and consume vast amounts of energy in the effort. Nature flies much more efficiently than technology.

One of the great hopes for our energy future is solar power, and we have applied our technological gifts to the design and manufacture of contraptions that can convert sunlight into electricity. But even the best minds on the planet, with all the resources of modern science at their disposal, have yet to develop anything remotely as efficient as the lowliest plant. Plants are incredible solar factories that convert sunlight all day, every day, in cold places and hot, wet and dry; and they do it with virtually no waste and in complete silence. Nature captures energy with breathtaking efficiency.

There are times, indeed, when nature seems perfect: a cheetah in pursuit, an orchid flower receiving a bee, a towering redwood, or a perfect tomato ripening on the vine. One response to this apparent perfection has been to conclude that nature is designed, the work of a creator. The truth is much more inspiring, of course: this world of apparent perfection arose on its own, with no direction whatsoever, as the myriad consequences of evolution.

Natural selection constantly hones living things to the utmost efficiency within their environment. The speed of the cheetah is repeatedly tested by the prey that attempts to outrun it; the orchid flower is repeatedly tweaked to ensure that the right pollinators will continue to visit; the virulence of the flu virus is tested by the immune systems of its hosts; and the ripe tomato is constantly perfected by the fruit eaters to whom its sugary redness says, "Eat me, disperse me, fertilize me." It is those with the most successful arrangement of traits—the fittest—that pass on their genes most abundantly to the candidates of the next generation. Natural selection, the great machine of efficiency, has spent four billion years sorting the strong from the weak, the fast from the slow, and the productive from the unproductive. But natural selection is only part of the story of life. The other part of the story is sex.

Sex is expensive, it is risky, and it requires two organisms—often three, in fact—to get together. It is a stunningly inefficient waste of time and energy, and it is extraordinarily dangerous. Most importantly, of course, in order to perform the fundamental task of transporting genes from one generation to the next, sex is completely unnecessary. And yet nearly all the big organisms on the planet go to great pains to couple.

As far as the expense of sex is concerned, consider the leatherback turtle. Here is an incredible animal, a relic of the age of the dinosaurs that grows to around seven feet in length and weighs in at six hundred pounds or more. It glides through the oceans all over the world, snapping up squid in the cold, deep waters from Greenland to Antarctica. But every summer, mature female leatherbacks do the oddest thing. They swim thousands of miles back to the tropics, to the very beach where they were born, and haul their cumbersome backsides inch by clumsy inch up onto the sand. They then dig a three-foot-deep hole in the sand with their flippers and lay their eggs. At the end of the laying period, their job done, they swim back to the other side of the world. That's a very expensive trip.

We could keep the discussion close to home and consider the expense of human sex. The best things in life are free? Not so much. We could monetize sex by looking at the costs of prostitutes,[4] for example, or the revenues of the porn industry,[5] but that would not apply to many of us (right?). If that's all a little sordid, unsatisfying, or expensive for you, consider the costs of finding a nice, reliable person with whom you can settle down and have sex, regularly, for free. Even if you're as cheap as me you'll have to invest in some half-decent clothes and a few movie tickets to pull that one off.

As far as the dangers of sex are concerned, consider the peacock. Here is a bird that must be a very tasty morsel for any number of predators in the forest: so much so that it spends much of its life looking skittishly over its shoulder to avoid getting eaten. But at breeding time, the peacock throws all caution to the wind, grows an immense rosette of brilliantly colored plumage, and struts around the forest like it's invincible: the cock of the walk, as we say in Manchester. That's a lot of risk to take on just to get laid. And the peacock is hardly alone. The males of many bird species come in distinctly predator-ready packages. Or consider the male praying mantis. He fusses around, finds a suitable mate, does his best by her, and then how does she repay him? Right. She bites his head off. What about elephant sex? Eek. That can't be risk free. Or porcupine sex: ouch.

Sexually transmitted diseases are a particularly unpleasant reminder that sex is dangerous, and many microbes have evolved to fill this unfortunate niche. They find their way from body to body during sex. Since their transmission requires sexually active animals, they tend to have a latent phase during which their symptoms do not appear, and they can hop, secretively, from individual to individual. Many sexually transmitted diseases only cause symptoms after a long incubation period, and syphilis is a particularly nasty example of this.[6] Syphilis causes nasty-looking sores but, just as you're thinking it may be time to head to the clinic, the sores disappear. Relieved, you go on about your business, figuring all is well. All is not well. The syphilis is now moving into your nervous system and other organs. One of the strange symptoms of the disease is periods of what might be described as "enhanced enlightenment." Deborah Hayden, in her book *Pox: Genius, Madness, and the Mysteries of Syphilis*,[7] contends that many brilliant artists raised their game in syphilis-induced neurological states. This state seems to bring on a range of odd behaviors, many of them risky, such as having lots more sex. This, of course, may be exactly what

Sex is expensive and risky. This snappy dude is looking for a girlfriend, but he's also making life much easier for the predators that are looking for him. It would be much more efficient to do away with this exotic display, but abandoning sex is a sure way to eventual extinction. *Courtesy of Steve Hallett.*

the syphilis wants. Eventually, the disease results in dementia and death. Not good. Wear a condom, kids, wear a condom: sex can be dangerous.

As far as the logistical problems presented by sex, consider the plants. Here are organisms that don't move much, so their specialty is the three-some. They coerce all manner of beasts—mostly flying ones—to serve as go-betweens. The orchid family is particularly fantastic in this regard. Orchids use highly specialized couriers to transport their sex cells directly from the boy orchid organs on one flower to the girl orchid organs on another. The orchid flowers have all their glorious shapes and colors to attract not the human eye but the insect one. The insect reaches for the nectar; the orchid exchanges its gametes. A South American orchid gives its sex couriers, small male bees, a sample of perfume that they use to attract members of the opposite sex. A tiny Australian orchid is even trickier. It produces a flower that looks like a female wasp. The boy wasp tries to mate with the fake female, hopping from flower to flower in a state of sexual frustration, spreading the flower's pollen as he goes.[8] Why do the plants go to all this trouble, we might ask, when the quack-grass rhizome can spread out in all directions, popping up new quack-grass plants everywhere, and the dandelion can spit out a shower of asexual seed?

The most important contradiction of sex is not that it is expensive, risky, and a logistical nightmare, but that it seems to be totally unnecessary. The bacteria skip the whole annoyance of sex altogether, reproducing perfectly well by making simple copies of themselves. They pass from generation to generation quite calmly. Why all the fuss? What is sex? Why didn't the efficiency machine of natural selection do away with sex long ago?

The answer is well known, of course. Sex is required to provide genetic variability. (This is why the bacteria don't need sex—they reproduce so often that the minor mistakes they make copying their DNA are enough to generate the required variability.[9]) But what does "producing genetic variability" really mean? Well, it means that while natural selection is busying itself with the task of perfecting organisms for efficiency, sex is doing the exact opposite. It is busying itself with screwing them up.

The fact that sex is essential to long-term survival on this planet is self-evident. Any species that abandoned sex must have been eventually whisked into extinction. The few oddballs out there that have abandoned sex have done so very recently, and they are doomed. Sex, then, is a system that shows phenomenal evolutionary foresight—it plans for the future—but that makes no sense either. Evolution has no foresight whatsoever. It operates strictly on a generation-to-generation basis and has no idea what traits will be most successful in a thousand generations.

Sex is best seen as a regulating device that tempers the tendency of natural selection to ensnare species in an efficiency trap. Unregulated, natural selection would repeatedly drive organisms to the highest level of efficiency, but it would also drive them to the highest level of simplicity. The most efficient organism, a success in its environment, would out-compete less efficient individuals, and natural selection would drive the less efficient organisms into extinction. Eventually, the most efficient individuals would be producing identical superefficient clones of themselves, and these offspring would be successful, too, but only while the environment remained unchanged.

Superefficient clones would benefit from the energy savings of abandoning sex in the short term, but they could never last.

Changes in the environment, such as decreased rainfall, increased temperatures, or lower humidity, might leave an organism high and dry in a place to which it is no longer well adapted. Even more dangerous would be changes in an organism's food supply or adaptation in its predators and parasites. Unable to respond to changing threats, superefficient clones would lose their edge and find themselves vulnerable. No longer able to evolve apace with the rest of the biological world, their eventual demise would be assured.

More than a billion years after sex first evolved, it remains a requirement for long-term survival. It is expensive, risky, and cumbersome, but life on earth has found through bitter experience that the surest path to extinction is to abstain from sex and fall into the efficiency trap. The antidote for efficiency in the natural world is sex. We need to find antidotes to efficiency in the modern world that work like sex.

The Self-Assembling World

To understand evolution is to have a breathtaking window onto the world. It's ironic when creationists look down on evolutionists as people who have given up on the world's mysteries. The opposite is true. To believe that a creator—an intelligent designer—came up with all this diversity is to accept an impossibly simplistic and banal explanation that quashes mystery. The realization that life arose and evolved with no outside help is adequately inspiring for me, thank you very much, and the realization that all this complexity, beauty, and diversity arose from a completely self-organizing system inspires enough awe for a lifetime. The constant interplay between efficiency-driven natural selection and its sexual counterbalance has built a phenomenally diverse biological world.

And life on earth is not only self-assembling, it's also self-regulating. The constant pressure of natural selection coupled with the constant meddling of sex has resulted in the evolution of a vast diversity of organisms that play out their great game of life on the canvas of planet earth. They interact as predators and prey, parasites and hosts, mutualists and competitors, and the myriad interactions among organisms and their environments present a dizzying cacophony of complexity. But, amid the confusion of a thousand interactions among fighting, cooperating, competing organisms that act in their own self-interest, an amazing degree of stability emerges.

Individual organisms may be fickle, and their fates may be unpredictable, but the communities and ecosystems in which they live have enormous resilience and stability. Despite the clamor within, ecosystems change only gradually, and they change in rather predictable ways. It is the complexity itself, and the interactions among a wide diversity of organisms, from which this stability is derived.

An overabundance of gazelles on the savanna is corrected by the lions that hunt them—but overzealous lions will be met with a shortage of gazelles. Lion and gazelle numbers are in feedback with one another, and their balance on the savanna is maintained. Squirrels buzz around, collecting the precious fruits of the oak tree, eating some and burying others for winter storage. Their prodigious efforts appear to vacuum the forest floor clean, but squirrels are exactly what the oaks need. The few acorns that are buried but lost are perfectly planted. New saplings emerge and the place of the oak as the king of trees is assured.

An ecosystem might seem to go through dramatic changes, but this, too, reveals an underlying resilience and predictability. When a wildfire sweeps through California's chaparral or Montana's Bitterroot Valley it appears as though a forest has suddenly—and tragically—been lost. Nothing could be further from the truth. Many forests need to burn if they are to retain their character. Fire does not destroy these forests, but renews them. Neither the mature forest, full of tall, green trees, nor the burned landscape of smoking stumps is a more important component of the ecosystem: they are parts of a dynamic, self-regulating whole.

Living systems teach us just how stable self-assembling systems can be. They generate a suite of complex and robust feedbacks and filters that maintain a system that appears impossibly random but is in surprisingly tight balance. The rules that are imposed are governed by the laws of nature, and the regulations, checks, and balances are those evolved over millennia of trial and error.

Human systems have a lot to learn from natural ones because we find it extremely difficult to develop systems that remain in balance in the long term: we tend to get caught in the efficiency trap. Human systems possess many of the efficiency-honing attributes of natural selection but few of the attributes of diversity-generating sexual reproduction. Our ability to extract natural resources, to compete, to develop new technologies, and to streamline our business develops apace: we become increasingly efficient. But our resilience erodes. We lack checks and balances; we lose diversity and robustness; we find ourselves perfectly adapted to the world as it is. But what if the world should change?

Why Smokey the Bear Causes Wildfires

> Beneath the zeal for efficiency lies the desire to control a
> changing world, by bringing it into conformity with a vision of
> how the world does or should work.
> —Jennifer K. Alexander, *The Mantra of Efficiency*[10]

Forests are complex ecosystems that come in many shapes and sizes, from wet tropical rainforest to dry chaparral. Some seldom burn and are seriously damaged by fire, but most burn at least occasionally and rebound from fire as part of a natural cycle. Many forests are deeply adapted to fire and need to burn periodically if they are to survive. That fire plays an important role in the ecology of forests, and has done so for a very long time, is evidenced by the many adaptations to fire that have evolved in plants. Ponderosa pines rarely suffer crown fires because they tend to shed their lower branches, restricting the penetration of ground fires into the canopy. Many eucalypts contain energy reserves in their roots, enabling them to resprout rapidly, even if most of the tree has burned. The cones of the lodgepole pine are sealed with a thick layer of resin that releases seeds when melted by the heat of a fire so that the seeds fall on fire-cleared, open ground.

Perhaps the most fascinating fire-adapted plant is Australia's mighty mountain ash, which happens to be one of the tallest trees in the world.[11] Mountain-ash forests go through the most fascinating fire cycle. The mature forest has a relatively open canopy that tends to support a rather dense understory of shrubs and ferns, including tall tree ferns. The seeds of the mountain ash fall beneath this dense shrub layer, where there is insufficient light for them to grow, so they lie dormant in the soil. The seeds cannot germinate until the tree ferns that block out the light are removed, which they are, eventually, by fire. When fire comes to the mountain-ash forest it can be a truly dramatic event. The trees are loaded with oils and burn extremely hot. Fire rips through the forest, razing it, and the seeds of the mountain ash are freed. Volatile compounds generated by the fire break the dormancy of the seeds, light floods the newly cleared ground, and the soil is fertilized with ash. The mountain ash seedlings burst to life, grow rapidly, and a new stand of saplings is established before the shrub layer can move back in to blanket the ground.

Fire is essential to the mountain ash. If the forest were to last for more than a couple of centuries without fire, the mountain ash's seeds would eventually decay in the soil; the mature trees above would die of old age and the mountain-ash forest would be gone. Is this why the trees are packed with so much oil? It seems paradoxical, but the mountain ash seems to have evolved to burst into flames. Its offspring are waiting patiently in the soil for the chance to start the next generation, and so the parents go out in a blaze of glory to give the offspring the best possible start in life.

The mountain-ash forest is an extreme example of an ecosystem shaped by fire, but fire is no stranger to forests. As forests mature, branches and twigs accumulate on the forest floor and the forest becomes gradually more combustible. Once sufficient fuel has built up, fire may come. If fire comes early it merely cuts through the understory and cleans it out. The trees may hardly be scorched. Here is where the erstwhile Smokey can get in the way. By preventing low-grade fires, Smokey allows forests to age unnaturally. He tries to prevent fire, but fire is a force that cannot be stopped forever—it can only be delayed.

Smokey is symptomatic of a culture that wants to sustain systems but, by failing to understand their dynamic nature, threatens them. Fire is no threat to unmanaged forests: it is part of a natural cycle. Fire prevention can break that cycle and, ironically, promote its own worst enemy: wildfire.

"Only you can prevent forest fires." Right, but if you keep putting out the small fires you may end up setting the conditions for huge wildfires. *Courtesy of Steve Hallett.*

One of the watershed events in forest management came in America's iconic Yellowstone National Park where fire had been suppressed for decades. Following a number of years of relatively moist, cool conditions came the hot, dry summer of 1988. A number of forest fires ignited, mostly outside the park, and coalesced into a number of very large blazes. More than nine thousand firefighters and four thousand military were called in to fight the fires and, by the end of the summer, when the cool fall weather finally put out the last of the fires, nearly eight hundred thousand acres of Yellowstone had burned: roughly a third of the park. The media converged on the park in a frenzy,[12] proclaiming the end of Yellowstone and that Yellowstone was burning down.

My daughter Sam and I can attest to the fact that Yellowstone remains one of the most breathtaking places in the world: we spent a fantastic vacation there during the summer of 2012 and, in just one week, saw virtually the full range of America's glorious native wildlife.[13]

Smokey the Bear was introduced by the forest service in the 1940s; part of his appeal was probably the need for wood during the Second World War and the intuitive sense that a burned forest represented wood lost to the war effort. The forest service, with Smokey as its mascot, had a huge impact on forest fires, reducing the total acreage of forest burned in the 1960s to about a tenth what it had been in the 1930s. By the 1980s huge numbers of fires had been prevented and the scene had been set for a rash of wildfires.

Yellowstone, ironically, had been one of the few places where park managers had a better understanding of the importance of fire in forest ecology. Rangers had been experimenting with modest controlled burns, but the dominant attitude to forest fires was still to put them out. The park service and the forest service had a different philosophy regarding fire in the 1980s, and this probably contributed to the confusion and media frenzy that surrounded the Yellowstone fires.

Significant change occurred in a single season and the acreage of mature forest was dramatically reduced, but a quarter century later, we still see a wonderful mosaic of different landscapes. Fire prevention leading up to 1988 certainly contributed to the burning of unusually large and furious fires in Yellowstone, but no permanent damage was done to its ecosystems. And why would it? Yellowstone is "the land of fire and ice" and always will be.

Thinking in Systems

Many different things can be considered systems. Biologists who study complex molecular and metabolic processes call themselves "systems biologists."[14] They might study photosynthesis, for example, a system of membranes, pigments, and enzymes that regulates the collection of carbon dioxide from the air and energy from sunlight, and the production of sugars in the chloroplast. Photosynthesis is a very complex system, and yet it is just one small part of the broader system of the plant cell that operates and regulates a dizzying array of other systems simultaneously. The cell, as complex as it may be, is just one part of the bigger plant system, and on it goes: the plant is part of a population that is part of a community, an ecological system in its own right that is part of an ecosystem, a big and complex system in which the biota of an environment interact. An

One of the first plants to colonize after forest fires is the aptly named fireweed, whose purple flowers soon blanket the ground. *Courtesy of Steve Hallett.*

ecosystem is a component of a biome within the biosphere, which is, in turn, a component of the greater earth system.[15] Like I said: complex.

So what do we need to understand in order to understand the whole thing? Well, the more we understand, the better, but the key is to at least recognize that systems operate at different scales and interact in hierarchical layers. Changes in one system can have far-reaching consequences in others, especially the larger system in which they interact. Consider the ecology of plants that thrive in the hot deserts compared to those that thrive in the wet tropics. To study their interactions at the ecosystem level, it is illuminating to understand their rather different photosynthetic systems. Plants of the wet tropics often have an extra photosynthetic pathway, the C4 pathway, which primes photosynthesis for high output in an environment with lots of sunlight and water. Desert plants have a different mechanism, one that enables them to shut down part of their photosynthetic system during the heat of the day so that they can conserve water. Thus, the molecular-level system has profound effects on the landscape-level one.

One of the most startling things about biological systems, at both the molecular and ecosystem levels, is how astonishingly well regulated they are. Photosynthesis speeds up, slows down, stops, and restarts as the need arises. Leaf pores open up to allow carbon dioxide to enter the leaf when it is needed and close when it is not.[16] The various reactions of the photosynthetic machinery are controlled by one another so that the right concentration of metabolites is provided to each at the right time. The glucose produced by photosynthesis is shuttled off to generate energy or synthesize the building blocks of the plant.

The next cohort of trees is taking back the land among the charred stumps of their parents. *Courtesy of Steve Hallett.*

This patch of Yellowstone probably burned in the huge fires of 1988. Its productivity is increasing rapidly, and with very little litter on the ground, it is unlikely that it will burn at this stage. *Courtesy of Steve Hallett.*

Ecosystems, too, are startlingly well regulated. The sudden proliferation of a particular species of plant might seem to threaten the stability of the ecosystem by shading out other species, but such dominance seldom lasts long. The new abundance of this plant is likely to be met by pests and parasites, or a shortage of pollinators, and its abundance will be corrected. The ecosystem, like the molecular system, contains an array of checks and balances that maintain it in a surprisingly stable state.

Systems have structure, undergo transformations from one form to another, and are regulated by an array of feedbacks. At base, however, they are simply networks through which energy and resources flow. The overall stability of biological systems is particularly impressive because it evolves out of pure selfishness and seeming chaos. None of the community, population, organism, cell, or metabolic systems has any role in designing any of this, and none of them strives for stability, and yet the hierarchy of complex systems bundles together and remains stable with no outside control exerted upon it whatsoever.

The key to understanding the stability of ecosystems—and the instability of human systems—is to understand how energy and resources flow through them. We have a tendency to think that human systems operate differently from natural systems and that we can circumvent some of their rules and put them under our control. In some cases we can do this, but our attempts at control more often give a false sense of security. What we fail to realize is that the system that we have built, admire, and are attempting to control, is ultimately under the control of nature—and nature's rules are strictly enforced.

Controlfreakonomics[17]

> Rational elites . . . know everything there is to know about
> their self-contained technical or scientific worlds, but lack a
> broader perspective. They range from Marxist cadres to Jesuits,
> from Harvard MBAs to army staff officers. . . . They have a
> common underlying concern: how to get their particular system
> to function. Meanwhile . . . civilization becomes increasingly
> directionless and incomprehensible.
>
> —John Ralston Saul[18]

Understanding the importance of fire in the ecology of forests does not always prevent people from building homes in fire-prone areas, or city zoning committees from allowing them to do so, but it has made the dangers much more clear. The use of prescribed burning has become very common in Australia, for example, where wildfires have always caused significant property damage. The first significant prescribed burn I saw was on the eastern slopes of the Great Dividing Range, close to the city of Toowoomba. An upscale subdivision had been built right up to the edge of the forest, but the forest service had decided a burn was needed nonetheless. The fire was impressive, and flames licked up close to the gutters of the houses, but no property was damaged and the neighborhood was made safe. Without the prescribed burn the forest would soon have reached the point of no return and fire would eventually have destroyed it.

Prescribed burns can sometimes go awry, burning hotter than expected and spreading farther than planned, so we are legitimately somewhat leery of them, but I think a more important reason we shunned fire for forest management was overconfidence in our ability to exert control over nature. It was obvious enough to earlier generations that one should "fight fire with fire." The inhabitants of the grasslands of the world have managed their environment with fire for thousands of years, so when did we lose the knack? It was only when we thought we knew better and could control nature that we chose to ignore this rule.

Another problem with our attempt to control forests appears to stem from a failure to understand the natural cycles through which they go and the misguided belief that the only important stage of development of a forest is the mature forest of tall trees. To foresters, mature trees represent more board feet, and to developers, fire control means they can plant more houses on the forest margin. Neither foresters nor developers are particularly fond of fire.

The sense that ecological systems could be controlled has hardly been limited to forests. Another example is fisheries. It started to become apparent in the 1960s that the stocks of certain key fish species were in dangerous decline. It's really obvious now, of course, since populations of species such as the bluefin tuna, Atlantic cod, and others have collapsed, but, at first, we believed that we could arrest the decline of fisheries by imposing fishing quotas.[19] As long as fishing was reduced to a reasonable level so that fish populations were maintained above certain thresholds, fisheries biologists believed that both fish populations and active fishing fleets could be maintained. There are a number of reasons that this approach was a widespread failure. The fishing limits were generally far too generous, and when the limits were exceeded, enforcement was lax and ineffective, but the problem runs much deeper than that. It is another example of how we fail when we try to control nature within simplistic limits and try to keep a dynamic system in stasis.

The oceans, like the forests, are complex systems, and reducing the populations of selected fish species has impacts on the system at large. Most of the fish species that we harvest from the oceans are the larger, predatory fish, and when we remove them we have an impact not just on a species, but on a trophic level. Removing cod from the ocean changes trophic interactions that impact the relationships among other species both higher and lower in the food chain, and one effect has been the ascendance of less-desirable scavengers like dogfish. It has also caused the proliferation of lobster, bottom feeders that clean up in an ocean system now dominated by smaller prizes. And the cod are not recovering, it seems. The ocean system seems to have tipped into an alternate steady state into which the cod can no longer find their former place.

Overfishing has also been a major contributing factor to the collapse of coral reefs, and our meager attempts to rein in the harvest have generally been of little use because the reefs are being affronted by a combination of problems, including global climate change and pollution. Roger Bradbury declared, in an article in the *New York Times*, that it's time we stop calling the coral reefs of the world endangered and admit that most of them are irreversibly doomed. He scoffed at the "consensus statement" from a recent coral-reef conference that called on "all governments to ensure the future of coral reefs." Too late, he says, "conservationists apparently value hope over truth."[20]

This desire to exert control over nature, and the belief that we can do so, seems to be a common human failing, and it is perhaps in agriculture that it reaches its height. Agriculture is an exercise in applied ecology in which we manipulate nature to produce an abundance of selected plants and animals. There is nothing wrong with this, but we do have a tendency to take it to extremes.

One of the most poignant examples of humankind's desire to conquer nature was the establishment of cotton production on the rivers feeding the Aral Sea in the former Soviet Union.[21] The Amu and Syr Rivers were diverted into hundreds of miles of canals that fed into hundreds of thousands of miles of on-farm irrigation channels. Trouble came extremely fast, and the sea began to dry up as its waters were diverted to the land. The fourth-largest freshwater lake in the world was removed in half a century in order to grow cotton. Not only was the sea drained, but an ecosystem and valuable fishery was ruined. The region's weather has now also become drier, which means, of course, that the cotton fields need even more water.

What kind of hubris is this? Is it the folly of the environmentally blind? Surprisingly, in this case, the Soviets apparently knew that the cotton production would drain the sea but went ahead with the plan nonetheless. The

Uzbeks and Kazakhs, who have lost their sea, continue to support the irriga-tion on the farmlands because the loss of the irrigation water would destroy the region's now-established economy.[22]

Another depressing story of cotton hubris comes from Australia's tropical north, where an irrigation system was set up around the Ord River. In this case, the production of cotton became profitable when new insecticides that could control the devastating Heliothis moth hit the market.[23] The booming cotton yields appeared to represent yet another example of humankind's suc-cessful conquest of nature until the pests evolved resistance to the insecticide and the cotton yields collapsed.

The basis of agriculture is not complicated. Adequate and reliable sources of sunlight, water, and nutrients are required to grow crops. The amount of sunlight a given region enjoys is more or less out of our control,[24] and sunlight cannot be exhausted, but nutrient inputs and irrigation water are a very dif-ferent story. Rivers, lakes, and underground aquifers can be depleted and soils can be degraded.

Our disastrous attempts to control natural systems are mirrored in simi-larly disastrous attempts to control social and economic systems. Believing that we can control natural systems to the extent that we try is arrogant. They are not ours to control and they contain much more complexity than we acknowledge. Social and economic systems are our creations, however, and they tend to be somewhat less complex. Perhaps we can have more success here? Past experience would suggest no such thing, of course, because the emergence, dominance, and disappearance of businesses, the booms and busts of the stock markets, and the rise and fall of civilizations are eerily reminiscent of the establishment, maturation, and combustion of the fire-prone forest.

Systems, whether they are natural or human-made, seem to have this tendency to cycle. We have learned—to a certain degree—to appreciate the importance of fire in the renewal of forest systems that have begun to accu-mulate deadwood and have reached their limits to growth, but we still fear the renewal of social and economic systems that have reached their limits. Renewal in economic systems can wipe out wealth in a short period of time. Renewal in social systems might come in the form of revolution.

Economists have made some inroads to understanding the cycling of markets, particularly in response to the Wall Street crash of 1929 and the Great Depression that followed, but I think most economists are still working from a tired old playbook that makes a number of catastrophic assumptions. First, their analysis of market cycles resembles the analyses of early forest ecologists in that, although they have made progress in recognizing the cyclic nature of their systems, their focus is not on figuring out how to manage cycles but rather on how to stop them. Economists are determined to promote economic growth above all else. They do not give serious consideration to the prospect that growth may not always be possible. They also continue to deny the central role that energy plays in the promotion of economic growth, especially the enormous growth we have seen in the last century. The availability of abundant, cheap energy is assumed, and so policies for a post-peak world have not been developed.

The insistence that our economic systems must grow in a world in which the abundance of resources is on the decline compels us to squeeze more out of the resources we have: the efficiency solution. As long as we assume that any economic downturn must be only temporary and that a return to growth is not only inevitable but also the paramount goal, we will continue to push for greater efficiency.

Our relentless quest to keep our economies growing against a backdrop of an overfull world with declining resources is a very big gamble, and trying to buttress growth in this way may be fatal. The conventional wisdom is that efficiency strengthens systems, but our attempts to control complex systems might be the very thing that weakens them the most.

Productivity, Connectivity, and Resilience

> Each climax formation is able to reproduce itself, repeating with
> essential fidelity the stages of its development. The life-history
> of a formation is a complex but definite process, comparable in
> its chief features with the life-history of an individual plant.
> —Frederic Clements, *Plant Succession:*
> *An Analysis of the Development of Vegetation*[25]

The most straightforward measure of a system is its productivity. How much standing timber is there in the forest? How many fish can we harvest from the ocean? How much cotton is the land yielding? Are a company's shares increasing in value? At what rate is the economy growing? As long as productivity is increasing we tend to conclude that all is well, and we tend to manage systems so that they will continue to produce. We extinguish fire to maintain the forest, set fish quotas to save the fishing industry, develop new pesticides and biotechnologies and apply fertilizer and irrigation water to maintain crop yields, streamline and cut out inefficiencies to maintain the bottom line, and stimulate the economy in an attempt to keep it in a state of perpetual growth. But the forests still burn, and hotter. Fisheries collapse and refuse to recover. Cotton yields plummet and the depleted land is abandoned. Economies still crash despite the lessons supposedly learned.

Perhaps the oddest thing is that we always seem surprised. The system seemed fine. It was still producing, still yielding, still growing. What went wrong?

Are there better measures of the functioning of systems and ecosystems that might allow us to predict their failure and enable us to manage them more effectively? That a system is still producing today is clearly no guarantee that it will continue to do so tomorrow. In fact, productivity may not only be a poor measure of system stability, it may be the worst.

The productivity of systems tends to be slow at first, before gradually accelerating into a period of rapid growth. The rate of growth may begin to level off, but there is still some measure of growth in most systems when they suddenly crash. Productivity rises and falls in a startlingly nonlinear way. Our assumption that systems are healthy if they are growing is false, and this should be cause for real concern.

We need to find better ways of measuring the health of systems that might indicate when they become prone to failure. What is going on in systems that, despite their continued productivity, might presage collapse?

Plant systems go through a fascinating sequence of vegetation types known as a vegetation succession. A classic succession occurs on beach fronts where the shifting sands, disturbed by the wind and sorely lacking in nutrients, present an apparently unsurpassable challenge for plants. Not many plants can survive here, but a few have found their niche. The long, fibrous rhizomes of marram grass crawl out and establish a beachhead.[26] New shoots pop up but are soon buried by the shifting sand. This would sound the death knell for most plants, but the erstwhile marram grass resprouts from its underground reserves and pops back out. As it is swallowed by the sand again and again, the determined marram grass creates a legacy of layer upon layer of buried rhizomes. Eventually, it binds a tall mound of sand together: a sand dune. The sand dune changes the physical conditions of the beach, creating a barrier behind which more plants, now sheltered from the wind, can grow, and a new, more diverse plant community develops. These new plants leave their mark on the land by leaching nutrients into the sand and by dying in place. A few animals will be attracted, too, and they will also leave their own nutrient packages. Soil is slowly formed and, as it is, an even greater diversity of plants will make their homes. Left long enough, the beach will become a community of grasses and forbs and, eventually, perhaps a forest.

The nutrient-poor, shifting sands of Lake Michigan's Silver Beach present an environment in which few plants can survive, yet marram grass thrives. It puts out tough, fibrous rhizomes that pop up new shoots each time it is buried, eventually building a sand dune. Behind the dune, sheltered from the wind, a new plant community can form. *Courtesy of Ashley Holmes.*

Bare rock, recently spewed from the ground, won't support much life. There are a few things, however, such as these hardy lichens, that begin the slow but steady process of building the soil upon which more and more productive plant communities will develop. *Courtesy of Steve Hallett.*

A similar thing can happen after a volcano. The land may be left looking like a moonscape, completely bereft of life, but lichens and other super-tough organisms will creep in, bringing a smattering of organic materials. Mosses and other small plants will eke out an existence in cracks in the landscape and a soil will slowly form. Eventually, the vegetation and the animals will return, and the landscape will be restored.

Another classic succession can be observed on abandoned farmland. The farmer plows the land and plants her crops in a clean field so that they face as little competition from other plants as possible. When weeds emerge she rids the field of them as best she can. Farming is the practice of arresting vegetation succession, but the succession always wants to proceed. Stop weeding, and proceed it will. The field will soon be choked by ragweeds, pigweeds, barnyard grass, and crabgrass. Longer-lived weeds such as docks, ragworts, and thistles will follow, and they will be replaced, in their turn, by shrubs and short-lived trees. After a century or so the farm will be gone and the forest will have quietly returned.

Frederic Clements recognized the predictability of this process in his classic 1916 publication *Plant Succession*.[27] He called the first weedy arrivals pioneers, described a sequence of intermediate stages, or seres, which would follow, and described a final stage, the climax community. Plant ecologists have studied succession ever since, refining Clements's analysis and describing plant succession in many ecosystems. Plant successions have a good deal in common, irrespective of the ecosystem concerned.

One interesting characteristic of succession is the type of plants that colonize at different stages. The pioneers tend to be short-lived, fast-growing species that spit out masses of small seed. Ecologists call these plants ruderals, or r-selected species. They tend to be well suited to disturbed environments and make their living by rapidly establishing in new areas. But the r-selected plants, with their short stature and short lifespan, are not built for the long haul and are soon pushed aside by stronger competitors. The next arrivals have their moment in the sun, but they, too, are gradually replaced by yet more competitive and more long-lived plants. The plants that eventually dominate the climax community, called K-selected species, employ a strategy opposite to that of the ruderals. Their growth is less explosive but they eventually become much larger and more dominant. It takes some time for them to establish control over the land and outgrow their weedier, r-selected brethren, but when they do, they dominate the ecosystem and persist. Their root systems control most of the available nutrients and their canopies capture most of the incoming light. Space no longer exists for newcomers, and the r-selected plants can no longer get a foothold back on the land.

One of the most important trends in plant succession, then, is the gradual shift from r-selected plants to K-selected ones, and this is what we readily observe as we watch an abandoned cornfield reverting to forest. We can think of a developing ecosystem such as a forest as if it were a weedy child developing into a shrubby adolescent and a forested adult.

There is much more to the transition from childhood to adulthood than just getting bigger, and the same is true for the transition from weedy field

The weedy, r-selected ruderals, such as this erstwhile common mullein, will add yet more nutrients to the landscape, and the vegetation succession will proceed on its way. *Courtesy of Steve Hallett.*

---

to forest. The weeds, like children, grow wastefully and are somewhat disconnected from the larger world. The weeds' world is in the here and now. It is of little consequence that the farmer's plow may soon tear up the land, except reconfirm the wisdom of an avowed goal of living in the present. Weeds have lots of acquaintances but make very few firm friends. Their pollen tends to be blown on the wind or spread by whatever insect happens by, and their seeds are parachuted on a breeze or scattered aimlessly by a passing animal. The weeds are gradually pushed aside by more grownup plants that can plan for the future and interact with their community. These more serious plants form mutualisms with fungi in the soil and become increasingly flamboyant in their flowery displays, encouraging specialized insects to transport their precious pollen. The fruits of the forest improve, too, and the small, utilitarian seeds of the weeds are replaced by increasingly sumptuous feasts for fruit dispersers. The daycare center of selfish weeds yields to the high school of shrubs (more connected to the world, but often prickly) and then matures into a well-connected community governed by grownups.

Plant succession is a process of growth, but also of maturation, and the formation of relationships is key. The earliest stages of succession, the r-phase, contain little more than a random gaggle of opportunistic individuals doing their own thing. The later stages of succession, the K-phase, is characterized by much more complex webs of interaction. The diversity of species colonizing the land has increased significantly, and so have the strategies by which they make their way in the world. The K-phase becomes a deeply connected ecosystem. There is much more to the system than our simplistic view of its productivity. Much more fundamental, although harder to see, is its connectivity.

I am probably stretching the analogy of the developing forest as a developing person way too far, but there is one last, important way in which the forest is strikingly like a person: it grows old.

The human body can sometimes hang on for one hundred years or more, but things start to go wrong in midlife (trust me on this), and the body loses its resilience. It becomes harder to recover from a game of racquetball, or from an illness or injury. By the time we reach our sixth decade, our resilience is in serious decline. The aging forest loses its resilience as well, although it doesn't show it like a middle-aged man, and it begins to lose some of its connectedness. To the untrained eye, the forest of towering trees may appear to be at the peak of its powers, but its diversity is in decline. When the trees have dominated the forest for a while, few of the weedier plants and shrubs will remain. Indeed, there may even be just a few dominant tree species that have now monopolized the forest. They work as hard as they can to squeeze efficiency out of the forest and maintain their monopoly. The forest is at its efficient and productive best, but it is losing its diversity and connectedness. Its resilience is eroding and it is ready to burn.

Forests, and many other systems, cycle through four distinct phases: development, maturation, collapse, and renewal. These are the r, K, Ω, and α phases shown in the following figure, the resilience triangle. It is the maturation phase that we need to understand much better than we do because many of our social and economic systems are in this stage now, as is our civilization as a whole. We live in the mature forest. It seems strong and powerful, and it's still producing, but it is losing its connectedness, and it's vulnerable. We push it to produce and, since resources are tight, we clamor for efficiency gains to get this done, but its resilience is eroding. We believe that efficiency will keep it alive, but efficiency simplifies it, weakens it, and pushes it to the edge of collapse.

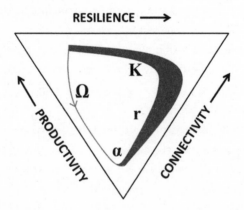

Cycles are common in natural and human systems. Systems grow wastefully at first (r), become more connected and productive (K), but can become overly mature and then collapse (Ω). They must then reorganize (α) before recovering. In the forest, weeds are replaced by shrubs and then trees that increasingly monopolize the land. Fire releases the monopoly, and the weeds reorganize. In business, entrepreneurs seize a new opportunity but become (and are bought out by) cumbersome companies that monopolize the marketplace. The idea grows old, and the entrepreneurs get back to work on something new. *Courtesy of Steve Hallett.*

Efficiency as a System Trap

Many systems cycle like the fire-prone forest, and they have different characteristics at each stage. Their growth is exponential in the r-phase because there are few limits on the availability of resources, and efficiency is generally low. High levels of productivity gradually increase the intensity of competition as resources decrease in relative abundance. The system increases its efficiency in the K-phase so that productivity can continue despite increasing resource limitations. Intense competition gradually winnows out competitors, and productivity is delivered by fewer, efficient members of the system that monopolize the access to resources. As it matures, the system, productive though it may seem, loses diversity and resilience. It is headed for the Ω-phase: release. Productivity crashes, resources are released, and the system must slowly reorganize through the α-phase before it can begin to grow again. This cycle is easily observed in many natural systems. I have used the example of the fire-prone forest but it is easily observed in grasslands, oceans, and many other ecosystems. It is also observed in many social and economic systems.

A new industry emerged in the early 1990s with the arrival of the Internet. Its emergence was probably best marked by the release of the Mosaic web browser in 1993. In its wake, a large number of entrepreneurs, reminiscent of weedy r-selected pioneer plants, set out to grow new businesses. Many would start the race, but relatively few would profit and survive. It's remarkable to realize how recent all this is. Amazon was launched in 1994, Yahoo and eBay in 1995, and Google in 1998. By the end of the decade almost any IT business was set to rake in investment, but investment was becoming disconnected from reality. Professional venture capitalists had begun the momentum, but bandwagon investing soon began and institutional investors and individuals put their money in these firms to avoid being left out in the cold. Many people had no idea what these companies had other than a ".com" suffix.

By the beginning of 2000 it should have been obvious that the competition was becoming tighter, the marketplace was filling with companies, and the resources they required—investment dollars and actual markets—were going to have to be earned. The scramble for resources began, and no fewer than seventeen Internet companies took out million-dollar Super Bowl ads in 2000. Publicly traded Internet companies were valued at $1.3 trillion even though many of them had yet to make a dime. On March 10, 2000, when the crash finally came, the NASDAQ, which had doubled its value to over five thousand that year, lost two-thirds of its value over the next.[28]

Dozens of companies failed within weeks. Cisco lost more than three-quarters of its value. Kibu.com, a company that had thrown a multimillion-dollar launch party just before the crash, survived in business for a grand total of forty-six days. The dot-com bubble released its resources like a forest fire, but, like the forest, the information-technology industry began to reorganize. Facebook, for example, was formed in 2004, during the recovery from the crash.

Well, thankfully, we learned from that little fiasco and our economy, as they love to say, was back on track. We know this, of course, because lots of people told us so. George W. Bush, for one, seemed to like the phrase "the fundamentals of our economy are sound" almost as much as "mission accomplished." The fundamentals were so sound, in fact, that there was no need to worry about the housing deadwood that was building up in the real-estate forest, there was no need to worry that the wealthiest 1 percent of Americans had monopolized so much of our wealth, and it was fine that a few big coal and petroleum companies dominated our energy markets—give 'em more tax breaks and subsidies, I say, it helps keep the weeds down.

The economic wildfire of 2008 is still with us. Thank goodness the economic forest service managed to hose it down enough to keep it producing. All we need to do now, it seems, is get our economies back on track, and that should not be too much of a problem because the fundamentals, I hear, are sound.

But let's hope there are no more fires.

CHAPTER 8
ALL THE OIL IN THE WORLD

We will consume all the coal, oil, and natural gas that we can.

—The author

No Good Deed Goes Unpunished

This global-climate-change mess we're in is really not our fault. The plants and microbes are to blame. We are the good guys, setting things straight.

Plants colonized the land during the Devonian era, about four hundred million years ago, when the planet was much warmer than today and the atmosphere had a much higher concentration of carbon dioxide. By the Carboniferous era, some fifty to one hundred million years later, some of these land plants had evolved huge stature. They were not the trees we are familiar with today; instead, they were giant lycophytes and ferns that formed huge, primordial forests. And a strange thing happened when they died.

Fallen trees are normally quickly consumed by microbes, and their nutrients, including their carbon, are restored to the biosphere. Huge numbers of trees from these primeval lycophyte forests, however, refused to decompose. The reason for this is generally considered to be that there were large areas of shallow ocean and swamps on the planet at that time, so the trees fell into water and their decomposition was slowed. Another theory is that the evolution of structural materials in plants, particularly the lignin that had enabled them to produce tall, strong trunks, had temporarily outpaced the evolution of microbes that could decay them. Plants had evolved lignin but the white rot fungi (now the preeminent global lignin decomposers) had not yet evolved the capacity to break it down.[1] Tree trunks, therefore, would lay around for much longer than they do today without decomposing and would eventually be buried and transformed into coal.[2] Massive volumes of carbon were taken out of the atmosphere and locked away.

Meanwhile, the high-carbon-dioxide, tropical conditions also caused massive microbial blooms in the oceans. Microbes died and sank by the trillions and accumulated in alcoves on the ocean floor, where they were slowly squished into oil and natural gas, removing still more carbon dioxide from the atmosphere.

The pesky plants and microbes reduced the carbon dioxide concentration of the atmosphere and cooled the climate dramatically. What were they thinking? Thank goodness we arrived to put things straight. We are finally returning that carbon, restoring the planet to a tropical paradise, and bringing the oceans back up to their proper level. We are the good guys, and this is not only an act of altruism, it's also an act of sacrifice. Unselfishly restoring the carbon balance of the planet, which was so egregiously put out of whack by those selfish plants and microbes, will destroy our civilization.

Conserve Oil: We Could, We Should, but We Won't

The unconventional wisdom places us face to face with the efficiency trap. Efficiency is hailed as a way of conserving energy but it does not. Even worse, the blind acceptance of efficiency can simplify and weaken systems that need to retain their diversity and resilience. The efficiency trap reveals that many of the goals that committed environmentalists are working diligently toward are futile. But people are not only failing to see the efficiency trap, they are failing even to ask the blindingly obvious first question that should come before it.

Conserve energy for what, exactly?

We are repeatedly told that we should consume less fossil fuel because this will make fossil fuels last longer and reduce the amount of carbon dioxide released into the atmosphere. It seems obvious that we should find ways to reduce our consumption of fossil fuels. Efficiency is not the answer, to be sure, but if you turn down the thermostat a little in the winter, wear a sweater, turn all the lights out every day, and cycle to work, you are conserving energy with limited rebound and staying, for the most part, outside the efficiency trap. That would seem to be progressive. Everybody agrees that actions such as these will reduce carbon dioxide emissions and save fossil fuels. But we are still not answering the first question.

Save them for what, exactly?

To answer this question we need to think about how much of the world's endowment of fossil fuels humankind will actually consume, and the answer to this question is simple: we will consume all the coal, oil, and natural gas that we can. This is a very important and troubling assumption, and it requires considerable reflection.

We know that reserves of coal, oil, and natural gas are finite. The amount of carbon stored underground that can be produced to generate energy—and release carbon dioxide—is therefore fixed. (And since we have reasonable geological data on fossil fuel reserves, the actual amount can also be estimated.[3])

If we want to prevent carbon from being released into the atmosphere as carbon dioxide, there is a simple method: leave it underground in its current form. No carbon sequestration technologies, hi-tech or low-tech, can ever be remotely as effective as leaving carbon in the ground in the form of coal, natural gas, and oil. If we were serious about preventing carbon dioxide from entering the atmosphere, this is what we would do: leave the carbon where the primordial forests hid it. Problem solved. But it's a preposterous idea, and there is no way in hell we are going to choose to do that. We will burn all the fossil fuels that we can. Again: the amount of fossil-fuel carbon underground is fixed, so the amount of carbon dioxide to be emitted into the atmosphere is fixed, too.

Leaving fossil fuels in the ground is what we feel as though we are doing when we make the switch from a Hummer to a Prius, replace the old furnace, turn down the thermostat, and cycle to work, but the saved carbon is still there to be taken, and taken it will be. Fossil fuels saved are not really saved at all. They are just saved to be burned another way on another day.

Perhaps we can leave carbon in the ground if we are able to replace fossil fuels with hydro, wind, solar, wave, geothermal, and nuclear? This possibility appeals to our optimistic natures, but realizing this possibility would require that alternative energy sources not only replace fossil fuels, a dubious prospect at best, but do so while fossil fuels can still be economically extracted. I doubt we could, but even if we could, I doubt we would.

I have worried about this problem at great length. I see no way out of it, and I think the assumption is correct. We will consume all the fossil fuels that we can, and we will only stop consuming them when we are no longer able to produce them. So what, exactly, is the value of saving fossil fuels?

Global Climate Change: Certainties and Uncertainties

If you are not up to speed on climate science, there are a number of excellent, accessible sources,[4] so I'll get through the basics very quickly. There are lots of things we know for a fact, lots of things we are fairly sure about, lots of concerns, and lots of uncertainties. The anti-climate-change lobby has been busy muddying the waters on climate information for a few decades now, so there's also a lot of disinformation. The disinformation tactics are actually quite revealing. Anti-climate-change activists attempted to dispute the science for a long time, but the science is increasingly secure and detailed. They have, as a result, increased their attacks on the scientists.[5] (If you don't like the message, shoot the messenger, but the facts are still the facts.)

The disputes about climate science, as far as I can make out, are not really driven by doubts about the science but about the politics and economics of the proposed solutions. Anti-climate-change lobbyists seem to have decided that, since they don't want to see action on global climate change, they should claim either that it doesn't exist or that it is not caused by humans. This approach is both logical and idiotic at the same time. It's logical because it's a good (albeit cynical) tactic, but their disregard for the facts is patently idiotic. They do have one good point though: the appropriate political and economic responses are not at all clear.

First, what we know for sure. We know that a number of important gases—the greenhouse gases[6]—trap heat in the troposphere and warm the planet. This is the greenhouse effect. We know that the atmospheric concentration of many of these gases has increased dramatically since the beginning of the Agricultural and Industrial Revolutions, and we know that human activities, including intensified farming, land degradation, deforestation, and the combustion of fossil fuels have been the primary cause. We know that global warming is happening, we know that it is not abating, and we know that it is anthropogenic (caused by humans).

Global warming is not the problem per se. The problem is the climate changes that come with it. We know a lot about what is going on here as well. We know that temperatures are increasing more rapidly at the poles than near the equator, and we know that regional weather patterns, especially precipitation patterns, are changing in many places, but we are less certain about many of the details. It is much easier to make broad, global predictions than specific, local ones. There is a reasonable level of certainty, for example, that grassland and semiarid regions, such as the Great Plains, the African Sahel, and large parts of Australia, will become hotter and drier, but what the weather will be in Oklahoma City, Timbuktu, and Alice Springs on June 21, 2050, cannot be known. Another example is severe weather such as hurricanes. It seems likely that global warming will increase the frequency and intensity of hurricanes, but there is considerable dispute about this (legitimate, scientific dispute, that is), and the details are hard to predict.[7]

This will suffice as a background here. There is going to be a lot of change, most of it bad and some of it catastrophic. The question for this book is, what should we do about it?

Various planet-modifying strategies, known as geoengineering, have been suggested, such as the placement of huge sunlight-reflecting mirrors in space or millions of tiny sunlight-reflecting particles in the atmosphere, but I'm going to ignore those here.[8] Instead, my focus will be on the greenhouse gases because that's where I think the answers need to be found.

The equation is relatively simple. The concentration of greenhouse gases in the atmosphere is a function of the amount emitted into the atmosphere and the amount captured from it. We have been emitting greenhouses gases by burning fossil fuels, among other things, and we have been reducing the capacity of the biosphere to capture greenhouse gases by deforestation and other acts of land degradation. The solution, therefore, if there is to be one, is to be found in reducing or reversing as much of this trend as possible: decrease emissions; increase capture.

This is where the political and economic arguments arise. Environmentalists and the political Left tend to support actions that they believe would decrease emissions and increase capture. The industrial lobby and the Right fear that such actions would threaten the economy and their entrenched positions so they do not support such actions and instead block them.

The argument centers on carbon dioxide because it is the greenhouse gas in greatest flux and over which we have had the greatest impact. The burning of fossil fuels converts vast quantities of solid, earthbound carbon into atmospheric carbon dioxide. Deforestation and land degradation inhibit the capture of atmospheric carbon dioxide for return to terra firma. There is an enormous amount of uncertainty about what we can and should do about global climate change, and the carbon wars rage on. Carbon exchanges have been formed to trade emissions reductions and carbon capture as commodities. Our good friend efficiency has been touted as essential to reducing emissions. Meetings of global leaders to address climate change come and go with murmured statements of concern, environmental piety, and zero action. A classic tragedy of the commons pervades,[9] exemplified by the question: Why should we use less oil if that will just let the Chinese get their hands on it?

We are completely missing the point and need to remember two ignored and misunderstood but pervasive facts.

Fossil fuels saved are only saved to be burned another way on another day: we will burn all the fossil fuels that we can.

Efficiency promises to conserve but actually consumes. Efficiency is a trap.

Is global climate change a huge threat? Yes. Is it anthropogenic? Yes. Are emissions of carbon dioxide from fossil fuels the main cause? Yes. Will fuel efficiency reduce emissions? No.

All the Oil We Can Burn, All the Carbon Dioxide We Can Emit

There is a good chance that you have not bought into the two previous statements, which argue that carbon emissions are inevitable and more or less fixed. If you're hearing this argument for the first time and do not doubt it, you're probably not really paying attention, because this is pretty heretical stuff. It bothered me for months. If we will consume all the fossil fuels we can, there is really no point in pretending to save them, and there's virtually nothing we can do to reduce the total amount of carbon dioxide we will emit. All we can do is either emit it a little faster or a little slower. And if efficiency increases rather than decreases consumption, then improved efficiency will cause us to emit it faster.

This way of thinking turns the global-climate-change question on its head. The question we normally ask is whether or not we can reduce emissions. Perhaps the question we should be asking is whether or not we can accelerate or delay emissions, but that is a fraught question, too. First, of course, we'd need to know which of these is preferable. It seems like a no-brainer, but it isn't.

Let's be clear: it's going to be disastrous either way. Climate change is bringing trouble no matter what we do, but do we want to slow it down or not? It seems obvious that we would want to slow it down, but would that just drag the pain out longer? I suspect delaying our carbon dioxide emissions would be desirable if we could manage it, and a delay might prevent the atmospheric carbon dioxide concentration from reaching its highest possible peak, perhaps somewhere around 700 ppm.[10]

As carbon dioxide concentrations continue to climb and the planet warms, a range of uncontrollable positive feedbacks will be—and are being—triggered. In the Arctic Ocean, for example, the melting of ice is accelerating warming as light-colored ice is replaced by dark-colored water that absorbs more energy from the sun. Also in the far North, melting permafrost will shed increasing volumes of heat-trapping methane. It would be lovely if we could avoid setting these feedbacks in motion because, once they are triggered, global warming will cause its own reinforcing acceleration.

Keeping carbon dioxide concentrations from reaching their highest possible level might be a lifesaver—but perhaps not. The lower peak would be maintained for longer (remember: the total volume is out of our control), and that's not an encouraging prospect either.

We are fumbling in the dark on the great challenge of our times. Global climate change is a runaway train heading for a missing bridge. We know that we are headed for disaster, and that we brought the disaster upon ourselves, but we are unable to react. We do not have the power to stop the train and, what's worse, everything we are trying to do to slow it down is causing it to accelerate. I am quite convinced that we are looking in the wrong places for mechanisms to counter global climate change. We continue to pretend that we can save fossil fuels, but we can't, and we have hitched our emissions-reduction wagon to the locomotive of efficiency.

Carbon Dioxide Is the Least Important Greenhouse Gas

Carbon dioxide has certainly been the major culprit of global climate change, and its concentration in the atmosphere has been increased by human activities since long before the Industrial Revolution.[11] In the centuries since the adoption of fossil fuels, our carbon dioxide emissions have increased rapidly and forests have fallen evermore rapidly to the axe (chainsaw). Carbon dioxide is certainly the most important cause of global climate change, but it might be difficult to do much about the atmospheric concentration of carbon dioxide. A few things would help, such as protecting and restoring forests and wild lands, and using strong regulations and emissions taxes to compel the coal industry to invest in carbon capture, but the carbon dioxide challenge might simply be intractable. The most important contribution to elevated carbon dioxide concentrations in the atmosphere is the combustion of fossil fuels, and we don't seem to have any way of preventing this voluntarily. The thing we are doing most actively is trying to improve our efficiency, but that is making matters worse. In place of focusing our efforts on carbon dioxide, perhaps we should turn our attention to greenhouse gases that we might be able to get under control. Although it is the major culprit, if we are looking for climate-change problems we can actually solve, carbon dioxide might be the least important greenhouse gas. Methane, nitrous oxide, and synthetic hydrocarbons might be better targets.[12]

The problem with synthetic hydrocarbons is a classic and depressing example of the first law of ecology: you can never do one thing.[13] One of the best examples of good environmental science combined with effective politics was the Montreal Protocol, which was signed in 1987. The treaty outlawed the use of CFCs in air conditioning and refrigeration because scientists had figured out that they were destroying the stratospheric ozone that protects us from damaging UV light.[14] CFCs were replaced by the rather similar but non-ozone-chomping HCFCs and HFCs. Unfortunately, although the ozone hole was a high priority problem in the 1980s, global climate change was not. CFCs are very nasty greenhouse gases, and it's great to be largely rid of them (and, yes, the ozone hole is slowly recovering),[15] but many HCFCs and HFCs are potent greenhouse gases, too. HFC-401a, labeled "environmentally friendly" because it was a non-ozone-depleting replacement for CFCs, causes over two thousand times the radiative forcing of carbon dioxide.

The air conditioning business is doing very well at the moment, boosted by economic growth in Asia, where air conditioner sales have increased by 20 percent per year for the last decade or so. Nearly three-quarters of the world's air conditioners are now produced in Asia.[16] The HCFC and HFC business is thriving as a result, but more of these gases are leaking into the air. The atmospheric concentrations of HCFCs and HFCs have doubled in the last twenty years and they are becoming a significant cause of global warming.[17]

Here might be a group of greenhouse gases that we can do something about. The HCFCs and HFCs that we currently use can be replaced with more benign compounds. The corporations will complain, as they always do, that the price of their units will become excessive, but it will not.

The next greenhouse gas upon which we should be focusing much more than we do is methane (natural gas). Methane has a radiative forcing twenty times that of carbon dioxide, and its concentration in the atmosphere has more than doubled since before the Agricultural and Industrial Revolutions. Methane is emitted from a strange collection of sources, and savings can be made at a number of them.

We have the oddest relationship with methane. So much so, in fact, that we give it two names. When it's fouling up the atmosphere or causing an explosion in a coal mine, we call it methane. When we want to use it as a fuel we call it natural gas. It's the same stuff: CH_4. Natural gas is the cleanest of our fossil fuels; it releases only half the carbon dioxide of coal when it burns, and very little of the other nasties that coal releases, but it's one of the worst greenhouse gases. Using natural gas is playing with fire. Natural gas burns cleaner than the other fossil fuels, but if we let it escape, it's very destructive, and we do let it escape in considerable volumes.

As much as 5 percent of our total production escapes from pipelines, especially the older cast iron pipes,[18] and the harder we push to produce natural gas, the more of it will escape to the atmosphere. Gas production has been holding steady in recent years, but it has been coming from far more wells drilled in less accessible places and from the hydraulic fracturing of gas shales. These tougher gas plays result in much more leakage. The cleanest of our fossil fuels is rapidly becoming much more damaging as we scavenge for it with greater desperation in much more challenging places.

Another source of methane emissions, and, amusingly enough, emissions is the perfect word in this case, is the backsides of cattle. Cows fart a lot, cow farts contain a lot of methane, and we now have a lot of cattle on the planet. Shocking, but true: the sky is filling with cow farts and the planet is getting warmer as a result. This particular brand of global warming has a very ancient history, dating back to the first pastoralists ten thousand years ago,[19] but it has certainly accelerated in the last century or so.

There are nearly a hundred million cattle in the United States today.[20] You probably haven't seen all that many of them because most of them are raised in confinement, but there's roughly one bovine out there for every three Americans. Why do we have so many cattle, you might wonder? The obvious answer is that we eat an awful lot of meat, and so a direct way to reduce methane emissions would be to eat less meat, but there is much more to it than that.

Meat is incredibly plentiful and cheap in America because corn is incredibly plentiful and cheap. Corn is the most efficient way of converting sunlight, fertilizer, and water into calories, and here is where we release large volumes of yet another greenhouse gas: nitrous oxide. We pour nitrogen fertilizer onto the land to produce corn and a fair bit of it is oxidized and released. Modern agriculture releases a truckload of greenhouse gases. Cattle fart methane and corn farts nitrous oxide. The Food and Agriculture Organization of the United

Nations (FAO) estimates that animal production rivals transportation for greenhouse gas emissions.[21]

Fixing American agriculture is a much more urgent and practical goal than trying to save fossil fuels, and doing so will not only reduce our emissions, it will also make our society more resilient in the process.

We seem to be spending a lot of our time pretending to fix global climate change, but our focus on carbon dioxide, especially our focus on fuel efficiency to reduce emissions, has become a distraction. Carbon dioxide, because we will have very little success in reducing its emission, is not the greenhouse gas upon which we should be focusing. Instead, we should be focusing on methane, nitrous oxide, and synthetic hydrocarbons.

Fossil Fuels Are the Least Important Resources

Questions of efficiency tend to focus on fossil fuels, and for obvious reasons. Fossil fuels are the wellspring of our modern world, and they are essential to our way of life. Their impending depletion is something of which we should all be afraid. Fossil fuels are also a major source of environmental pollutants such as lead and sulfur, and they are, of course, the major cause of global climate change. Fossil fuels bring both our greatest opportunities and our greatest problems, and so it is quite natural to focus our solutions on them, but fossil fuels are finite and we will have to live without them sooner or later. It looks like an impossible task, but the harsh reality is that we have no choice. And in any case, the method we have embraced for saving fossil fuels—that is, efficiency—will do nothing to help anyway.

Instead of focusing on saving things that we'll have to live without, we should focus on saving things we can't live without. It would be nice to still have clean air, fresh water, decent soils, forests, and some wild lands left when the fossil-fuel era reaches its inevitable end. (It would be nice to have a reasonable climate, too, but there's nothing much we can do about that.) And if we were intent on saving fossil fuels, we would be best served by restricting their use to production of chemical feedstocks for fertilizers, plastics, pharmaceuticals, and other chemicals. It never really made much sense to burn these valuable resources in the first place.

There are many resources that we will never see again. This surge of wealth that we call the modern world has also been a surge of environmental destruction. It has had such a huge impact on the earth's biota that it may be known to future geologists as the great anthropocene extinction event. The destruction of our coral reefs is a case in point. Most of them may be completely beyond recovery, representing the ongoing and impending extinction of large numbers of species, the death of entire ecosystems, and possibly the end of an entire biome. The importance of saving fossil fuels pales in comparison to the importance of saving coral reefs. The end of fossil fuels was always inevitable. The end of coral reefs was not.

The same is true for tropical rain forests, the destruction of which has expanded rapidly through the petroleum interval, and for our supplies of clean, fresh water and the availability of deep, fertile soils. When the petroleum interval comes to a close, as inevitably it must, we will find that we have used up not only a once-in-history endowment of powerful, versatile energy, but also many of the resources upon which we have relied for thousands of generations.

Given the choice of saving oil, natural gas, and coal compared to soil, water, and wood, the choice is simple. Let the oil burn. Let the forest stand.

Drill, Baby, Drill

I think the time has come to realign our focus on the whole issue of fossil fuels and global climate change. Massive carbon dioxide releases from the combustion of fossil fuels are changing the climate, and the disastrous legacies of this will haunt us for generations, but there's virtually nothing we can do about it. This colossus of a civilization that we have built on the back of fossil fuels is in great danger as the fossil fuels go into depletion, but there's nothing we can do about that either. In our desperation to curtail emissions and make our fossil fuels last longer we are crying out for efficiency improvements, but efficiency improvements will help not one iota, and improved efficiency will ultimately make matters worse. It's time to give up on most of these efforts and instead focus on surviving and recovering from the disasters that are about to visit us.

The only thing we can possibly do to reduce fossil-fuel emissions is to prevent the coal and petroleum industries from getting their hands on coal, oil, and natural gas. There is little hope for this, but we should, wherever we can, lobby against exploration, especially on the American Continental Shelf and in the Arctic National Wildlife Refuge. We should lobby against the Keystone XL Pipeline, which would expand production of the dirty Alberta tar sands and the development of shale-gas resources that cause groundwater contamination and will increase our methane emissions. Other than that, the best way to reduce carbon emissions from fossil fuels is to let the coal and petroleum industries deplete their resources as fast as they can so they don't have time to develop any more efficient technologies.

Our capacity to recover fossil fuels is affected by both technological and economic factors. Fossil fuels will be recovered for as long as it is economically viable to do so, and production will continue as long as there is profit in it. Increasingly efficient technologies reduce the costs of exploration, recovery, and end use. The combined efforts of our politicians, economists, and industrialists are working to ensure that fossil fuels can be recovered and used as cheaply and efficiently as possible. It is a deeply counterintuitive argument, but from a global-climate-change perspective, and from the perspective of our rapidly depleting resources, it would be better if they failed.

The fossil-fuel industries have gradually improved the efficiency with which they extract, refine, and deploy their products. The coal industry initially used human and animal power to pull the black rock from the ground, then it began to use steam engines, and now it uses massive diesel-powered machinery. These advances have enabled the extraction of coal from deeper mines. We even move mountains to get to it. The oil industry has used horizontal drilling, water injection, and a number of other techniques to gradually increase the proportion of oil that can be lifted from reservoirs, and it has figured out how to separate gas and oil from tight plays. It is very difficult to guess how much more productive and efficient these industries can become in the future, but it will depend on the interplay of economics and technology. Suitable economic conditions for a longer period of time, enabling more efficient technologies to be developed, will yield more fossil fuels. Most people would think this is what we want, but in the long run, it is not.

So the best possible solution is "drill, baby, drill"? Is it possible that those Tea Party Republicans are right, even if they're right for all the wrong reasons? Perhaps the best thing we can do with our fossil fuels is to burn them away as quickly and inefficiently as we can to make sure nobody puts them to "good use."

CHAPTER 9
EFFICIENT BUSINESS AND ECONOMY TRAPS

Promoting energy efficiency is an "obvious," common-sense way to reduce energy use. If roads are becoming more congested, the "obvious" response is to add capacity. If a bus or train service has lost so many passengers it is running at a loss, the "obvious" response is to cut back to smaller, less frequent buses, stopping earlier in the evening. But these responses stoke up the feedback loops. Often the "obvious" response simply makes the problem worse.

—Roger Levett, *Rebound and Rational Public Policy Making*[1]

Half my morning is gobbled up by e-mail these days. How did that happen? E-mail is easier and quicker than fax, phone, and mail. It is an incredibly efficient way of transmitting information, so shouldn't it save time? And I thought I was supposed to be saving paper in this electronic world. I do, in some ways, but I waste reams of paper writing on a computer. I could have written this entire book electronically and printed it off at the end, but word processing is such an efficient way of slicing and dicing the written word that I tend to print off a first draft that I know is pretty awful, edit it with pen and paper, and then do a rewrite. And then another. And then another.[2] I type with the accuracy of a hoofed animal, but that doesn't matter on an efficient computer. And it's not just me. We talk about the emergence of a paper-free world and yet global paper consumption has tripled over the last three decades. The efficiency gains of electronic media have increased the throughput of paper.[3]

Another random efficiency trap is seen with head injuries in football. It rather amuses me when Americans compare football with rugby, a somewhat similar game from the old country. "Those guys are crazy! Why on earth don't they wear helmets?" Both games are pretty violent and involve one big guy stopping another big guy by whatever means are available, sometimes the head. Football players started wearing fetching leather caps and then helmets to prevent head injuries, but while the number of head injuries in rugby stayed more or less the same, the number of head injuries in football rose. Do helmets, therefore, actually cause head injuries? The effect of helmets is to increase the speed of the game and broaden the types of blocks and tackles that are used, resulting in more injuries. The proposed solution, of course, is to make better helmets, but this makes the game faster still.

And why are racehorses shattering their cannon bones all the time on the track? We breed them for efficiency; to go faster and faster until their legs fail. Labrador retrievers tend to suffer hip dysplasia because we have bred them to have all those lovely lab qualities but overlooked the fact that we have simplified their genetics—inbred them—for dodgy hips. And, on the subject of efficient dog breeding, we now have the unfortunate "puppy mill" phenomenon that can efficiently churn out cheap canine Christmas presents as fast as we can dump unwanted mutts at the pound.

The Efficiency Trap is not a business book—lord knows there are enough of those out there—but if efficiency has become a mantra anywhere, it is in business, management, and economics, so we'll take a short detour into the world of business efficiency. Business books can teach you how to *Work a Four-Hour Day: Achieving Business Efficiency on Your Own Terms* (although I don't see why I should double the length of my work day), explain how we are *Thirsting for Efficiency*, and show us *The Pittsburgh Way to Efficient Healthcare: Improving Patient Care Using Toyota-Based Methods*. If that's all too highfalutin for you, check out *Business Efficiency for Dummies* or *Lean for Dummies*.[4] The titles I've picked here are actually pretty good, but much of the rest, and there's plenty of it, is just a morass of reruns of the conventional wisdom on efficiency.[5] But if efficiency is such a boon to business, and everyone is striving for efficiency, why do so many businesses fail? Perhaps they are not yet efficient enough and need the advice, but perhaps efficiency is not the solution they seek.

Just in Time, Going Lean

> Two things are going on at the same time with the flattening of
> the world: The relentless quest for efficiency is squeezing some
> of the fat out of life.
>
> —Thomas Friedman, *The World is Flat:*
> *A Brief History of the Twenty-First Century*[6]

James Womack, Daniel Jones, and Daniel Roos coined the term *lean production*
in their landmark 1990 book *The Machine That Changed the World*, in which they
described the ascendency of the Japanese automobile manufacturers and gave
us the first serious insight into the Toyota Production System.[7] Toyota had
become so effective, claimed the authors, because it had adopted an ethos of
lean production over mass production. Lean production is sometimes called
lean manufacturing, lean enterprise, or just plain "lean," and the idea has
spread into all manner of businesses. We now have lean construction, lean
higher education, lean healthcare, lean IT, lean dynamics, lean consumption,
lean startup, and more. The essence of "lean" is the stripping of nonessential
activities and avoiding waste. It could be summed up as preserving value with
less work. It's a classic efficiency measure: value per work.

The success of the Toyota Production System has made its founder, Taichi
Ohno, a real-life business guru, deified most completely in Jeffrey Liker's
2004 book *The Toyota Way: Fourteen Management Principles from the World's Greatest
Manufacturer*.[8] Ohno's ideas resonated with Western industries (actually more so
than Japanese ones) and I have a sneaking suspicion that part of the reason is
that its Japanese terms made relatively commonplace things sound exotic. The
idea of *muda* seems original, but muda is basically waste, and making muda is a
sin. *Hansei* is another term that, translated attractively, is the mystic-sounding
"purposeful reflection," but it basically means checking and analyzing progress. Trimming muda is a fairly standard efficiency idea, but enhancing hansei
goes a step beyond what manufacturers had previously considered, and it is an
efficiency trap avoidance strategy. It calls on manufacturers to seriously reflect
on the big picture of how the production system works.

Six Sigma was developed by Motorola in 1986, and its avowed purpose was to improve quality by improving the capacity of systems to identify defects. While fancy Japanese words added mystique to Toyota-style business systems, Six Sigma assigns people belt colors similar to those used in karate. A Six Sigma green belt has responsibilities for seeking out quality-system weaknesses. A Six Sigma black belt will kick your quality ass. Six Sigma also has a sort of quasi-mathematical, clever-sounding "statistical" underpinning, the sigma being a rating of the number of defects in a manufacturing process. Being six sigmas from the normal curve should assure high quality. (As fancy-sounding as it is, it's still an arbitrary benchmark, but never mind that.) Six Sigma is based around two methodologies designated by acronyms: DMAIC and DMADV. DMAIC, suggested as a route to improving established business, stresses the processes of Defining, Measuring, Analyzing, Improving, and Controlling the system. DMADV, a variant suggested for new businesses, stands for Define, Measure, Analyze, Design, and Verify. Six Sigma also stands out for having semiformal-ized statistical tools for analyzing the success of each step. Also in vogue is the attempt to combine the efficiencies of lean production with the quality-assurance strengths of Six Sigma. This is known as Lean Six Sigma.[9]

I'm not a business man but I find these approaches fascinating, and the fact that business strategies are attempting to employ some degree of systems thinking is extremely encouraging. But there are problems. I think the systems most businesses analyze are very narrow, and this leaves them vulnerable. Overconfidence in the belief that they are following an optimum systems approach might even leave them more vulnerable if they are not really on the right track, and I think many businesses have followed Six Sigma–type approaches in a rather one-size-fits-all way, which is not really much use. Such companies are paying lip service to systems analysis but are not really thinking outside the box.

The more encouraging aspects of approaches such as Six Sigma are those that encourage the type of analysis that can enable companies to adapt quickly. Just as in the biological world, it is not the strongest or fastest that survives, nor the most productive or efficient, but the one that is most able to adapt. Many of the other aspects of Six Sigma–type approaches are more troubling, particularly those related to streamlining through efficiency.

Another efficiency trend in business has been "just in time" production (which has its Japanese term, of course: *kanban*). Again, this is an important systems approach that attempts to optimize the flow of goods or services. In automobile manufacturing, for example, it might concern the optimization of the delivery of parts to an assembly line. The delivery of too few wheels would obviously interrupt assembly, but the delivery of too many wheels is also inefficient. It creates *muda* because the surplus wheels would require storage space and extra time and money would be spent moving them around.

Just in time, however, bears some risks. If it is too finely optimized, an unexpected interruption in any part of the system can jam the whole thing. Toyota faced huge problems after the earthquake, tsunami, and ensuing power shortages in 2011, and its problems were exacerbated because production disruptions backed up its system very quickly (although this is a rather extreme case). Another example was that of Ericsson, the cell-phone manufacturer that made the mistake of putting all its chips in one efficient basket: a production facility in New Mexico. A lightning strike in 2000 put the chip producer offline and Ericsson's losses forced it to join with Sony the next year.

Companies are going to need some serious systems thinking as we move into the next few decades because the playing field is going to change significantly. I wonder how many are considering this, and I wonder which will be able to adapt to the enormous challenges of economic stagnation and energy shortages that they will face. Very broadly speaking, I think those that are too reliant on lean and just-in-time systems will be in considerable trouble. Now might be the time to soften that approach and think about having more inventory on hand. Squeezing tighter and tighter for efficiency will eventually lead to a dead end. Those that think big, predict how changes will affect their business model, and adapt will fare better. Adaptation, for many companies, will involve downsizing, and this is not something that tends to come easily. The most important question for businesses going forward is, are you in a position to transition? Much more emphasis should be placed on alternatives, contingency plans, and business models that build robustness and resilience rather than efficiency because the business environment is poised to change dramatically.

Why the Biggest Employer in America Causes Unemployment[10]

Walmart is a colossal company that has been an exemplar of lean and just-in-time systems. It is a huge company that has had huge impacts on the world. Walmart is a great microcosm of how the globalized world works and an extremely good focus for an analysis of where the world might be headed. From the standpoint of discussing efficiency, there is no better lab rat. The Walmart system is the embodiment of efficiency at every step. Goods and services are obtained with huge economies of scale from wherever in the world it is most efficient to do so. The economies are then mobilized en masse. Walmart runs as lean as it possibly can, reducing inventory by using trucks as if they were just-in-time rolling warehouses. Walmart stores are lean and mean as well. You don't go there for service, but you'll find a vast array of products. Even more important than the range of products is the price. Walmart gives good value, there's no question about it, and they never have sales because one of their most important business tactics is to keep prices as low as possible even if it pushes down their margins.

Always getting bigger, increasing economies of scale, squeezing efficiencies out of every process, and selling at the lowest possible price has led Walmart to global retail dominance. It has also had a profound effect on society, which Charles Fishman calls the Walmart Effect.[11] Procuring goods and services at the lowest price does not sound all that intimidating, but it can be a big deal to be a supplier to Walmart. Having such a huge client can make your company—but it can also break it. Walmart inserts itself into many of its suppliers' businesses, and it has very stringent requirements for quality and on-time delivery. Get it wrong and you may very suddenly find you are no longer supplying to a client upon which you had become dependent. Many suppliers have modified their businesses to increase throughput and efficiency so that they can supply to Walmart—and then regretted it.

Walmart decided to bring people into their stores with a ridiculous offer of a gallon jar of dill pickles for $2.97. That's a monster-sized jar for virtually nothing. Now that doesn't seem like a big deal, but the gallon jars were all supplied by Vlasic, and they went into three thousand Walmart stores simultaneously. And they sold like crazy. Vlasic was churning through field after field full of cucumbers, had to start dropping other customers to fulfill their contract with Walmart, and were making less than two cents per jar. Vlasic filed for bankruptcy not long after the deal ended.

The most obvious Walmart Effect has been the gradual disappearance of small businesses. It's awfully hard for a business selling any of Walmart's product line to compete once the big retail monster comes to town. Walmart competes—brutally—with bike shops, bakeries, butchers, grocers, clothing stores, and plant nurseries, making them disappear. Walmart boasts that it is bringing employment when it comes to town, but the unskilled, low-paying jobs that it offers do not replace the jobs and small business owners that it ends.

The all-encompassing efficiency of Walmart has resulted in a massive and incredibly successful company, but it has had disastrous consequences for society. Walmart's efficiency reduces the diversity of our towns and cities, distorts our social landscape, and causes unemployment. So why do we keep shopping at Walmart? Well, because Walmart is efficient for us, too. Walmart is by far the best place to shop for the consumer seeking price efficiency and time efficiency. Society, then, has made the choice to trade quality and diversity for efficiency. But it was never really a choice, was it? It was an efficiency trap.

We have given away our mom-and-pop stores and lost the heart of our communities. We have moved increasingly to the suburbs from where we now need to commute. We have chosen to ship our food from across the country and the world, and buy our gadgets from China. As we move into the next decades we will regret having streamlined our societies in this way.

Understanding the growth of Walmart and its impacts on society is relatively straightforward, but what lies in the future for Walmart? Here is a much more difficult analysis, but I think there is trouble ahead. Walmart is so big and so powerful that it seems invulnerable, but Walmart faces threats at every turn, and part of the problem it faces is the size and scope it has achieved.

The temptation for Walmart will be to push the boundaries of size and efficiency like the Romans trying to defend an overgrown and bloated empire.

Walmart is deeply reliant on transportation in all its forms, shipping huge volumes of products from all over the country and the world, and this has been one of the backbones of the Walmart model. The costs saved by using cheap Chinese labor have outweighed the expenditures made in transportation, but that is going to change. Not only will the costs of transportation rise, but so might the costs of Chinese labor.

Perhaps Walmart's biggest vulnerability, however, is its laser focus on low prices. Walmart operates on extremely tight margins and fine-tuned efficiency, but the days of easy productivity increases are done and it's not clear that further efficiencies can keep the system alive. Walmart has two choices. Downsize and let some of the air out of the tires or keep pushing for efficiency to the end. The end, however, when it comes, could be quite spectacular.

Efficiency traps are everywhere in the business world and need to be understood. If we can understand the role of efficiency in a system as complex as an ecosystem, surely we can understand it in an economic or business system. Like many other systems, including ecosystems, businesses go through cycles. A new business opportunity is like a gap in the forest canopy; the entrepreneurs move in like weeds. Innovation is high, competition is low, and growth is rapid. Competition eventually begins and there are soon too many companies for the marketplace. Some companies fail and others merge to capture a larger share of the available resources for themselves. Mergers can result in monopolies, and what was once a new opportunity filled with diverse innovators becomes increasingly simplified and vulnerable. When the easy productivity gains are no longer available, the dominant companies reach more and more for efficiency to maintain a state of growth. They streamline their businesses, trimming the fat to maintain the bottom line, and then they file for bankruptcy.

The CCPP Game

> A comprehensive history of great business fortunes would
> show a disconcertingly large number that were made [where]
> the enterpriser devised a silent way to commonize costs while
> continuing to privatize the profits.
>
> —Garrett Hardin, *Filters against Folly*[12]

Rich countries have a secret that would be best kept from poor countries, and corporations share the same secret, which would be best kept from you. We are told about all the various secrets for success: efficiency, quality, adaptability, hiring and developing the best people, and so on. Sure, these are all important, but throughout the history of economics there has been one true secret to success: playing the CCPP game. Commonize the costs and privatize the profits.

The Walmart business strategy is a classic example of CCPP. It avoids the costs of labor and environmental pollution by sourcing overseas, drives down the price of goods with huge economies of scale, and then pockets the profits from its trades. Bear in mind that Walmart merely distributes goods from here to there, commonizing the costs wherever possible—it manufactures absolutely nothing.

A mining company is another perfect example of CCPP that privatizes profits by selling the coal or ore that it mines. The government gets a small piece in taxes (not nearly enough), but that's it. The company, meanwhile, commonizes a vast array of costs. It uses roads and energy subsidized by public funds, degrades land that could have been used by others or left wild for the common good, and pollutes the broader environment. The profits go to the company while the costs are shared among all. No wonder there is money in mining.

An oil company does the same thing. It degrades everybody's land, pollutes everybody's air, and collects the benefits of being in a developed society. Then it pockets the cash. A forestry company sells trees—but whose trees were they, and on whose land?

The CCPP game applies to all corporations. An IT company might seem to be more immune to it, but think of all the public services upon which a Microsoft or an Apple draws. They are huge beneficiaries of the publicly funded education system, the national electricity grid, and many other services. We should all demand a free iPad!

And what about those financial companies? I'm thinking of, oh, let's say AIG, Citibank, Bank of America, Morgan Stanley, JP Morgan Chase, for example. They benefit from lots of public services, take ridiculous financial gambles of the "I'll put a thousand on that three-legged horse" variety, pull in massive profits—for themselves—and then, when it all goes sour, take a piece of the half-trillion bailout and offload "toxic assets" onto the rest of us. This is CCPP at its finest.

It's not much of a stretch to see that CCPP is the body and soul of the global economy. The nations that have competed well against other nations are the ones that have privatized more of the profits from global resources, and shared the misery around.

The classic modern-world CCPP game is the oil industry. An American oil company doesn't just play CCPP at home and commonize its costs onto Americans; it plays CCPP brilliantly on the world stage. To do so, it uses vast amounts of public funds, the most obvious of which I call the "Pentagon Tax." It's a huge problem for rich countries that so much of the world's oil reserves are overseas, so the US government invests vast resources ensuring that American companies have access to the oil. We have assassinated, manipulated, sanctioned, and invaded to make sure oil remains on the "free" market. The United States is by far the world's biggest importer of oil, and yet it is also a huge exporter of refined oil products. This is classic CCPP. Import cheap oil; export expensive gasoline.

Cheap raw materials, medium-priced processed products, and high-priced finished products flow around the world, and a key to winning the CCPP game is, as always, to buy low and sell high. Selling off raw materials is a major waste of income and it also depletes your resource base. Buying finished products is expensive. Not surprisingly, rich countries are experts at buying low and selling high. Resources flow into the wealthy economies that have the capacity to add value to them. This is no surprise, of course, given that they make the rules.

Understanding the CCPP game makes it clear that an efficient economy might be something to keep an eye on if it means that someone—or some corporation—is efficiently taking your stuff to line their pockets, making you help with the clean-up, and then dragging you down with them.

The Myth of the Efficient Market

> Almost nobody (with the possible exception of a trained neoclassical economist) would doubt that energy is as essential to the functioning of the global economic systems as gasoline is to a car or electricity to a lightbulb.
> —Robert Ayres and Benjamin Warr, "Energy Efficiency and Economic Growth: The Rebound Effect as a Driver"[13]

The term *efficiency* is used a number of different ways by economists, and there are various iterations of economic efficiency. It can refer to the situation where one individual or entity profits without another making a loss, which is generally referred to as pareto efficiency, or it could refer to an economy that is running at maximum output on the resources available. It can also be a strict measure of the amount of economic output per unit of input. The input could be any number of things, but we will focus on energy, so economic efficiency, for our purposes, refers to economic output in GDP (gross domestic product) per unit energy. This is also referred to as the energy intensity of an economy.

Economists have come up with these various measures of economic efficiency so that they can compare different economic models or ideologies. (More cynically, you might argue that they have done it to support their own ideology, because the interpretations vary wildly.) The conventional wisdom is that the energy efficiency of modern free-market economies has improved over the last few decades. In other words, GDP has grown much faster than energy use. The energy efficiency of the US economy doubled between 1970 and 2008 because the amount of energy required to generate one dollar of GDP fell from 18,000 to 9,000 BTUs.[14] As usual, this is claimed as an energy saving, and since it's measured at the scale of an entire economy, it's a huge one. Another term for this is *decoupling*. The economy is said to have been decoupled from its demand on the environment because it has become so efficient. Economic growth has occurred without the need for more energy.

The idea of decoupling is linked to a broader concept known as the Kuznets curve,[15] which describes a U-shaped relationship between economic development and environmental protection. Extremely poor countries tend to have limited environmental impacts because they lack industries. Countries that are developing rapidly tend to have rapidly growing industrial sectors but poor environmental regulations and so they are the worst polluters. Fully mature, developed economies have the capacity to invest in environmental protection. The Kuznets curve has become a big favorite of free-market economists because it seems to show that the best route to environmental protection is through wealth. This lovely link shows us that we can have our economic cake and improve the environment when we eat it.

There is no such energy saving.

Did we forget what the guru of the Chicago school of economics, Milton Friedman himself, taught us? There is no such thing as a free lunch.[16]

First of all, the amount of energy consumed by the US economy increased by 30 percent between 1970 and 2010. Economists also seem to conveniently ignore the fact that much of the energy consumption of the United States has been exported. The United States and other Western economies have moved away from manufacturing toward service- and technology-based economies, but they still need steel, aluminum, and manufactured products. Instead of making these products at home, where the energy use and pollution would score against us, we have simply had them manufactured overseas. Not surprisingly, while energy use in the United States increased by 30 percent, energy use in China increased by 400 percent.[17]

An Efficient Global Economy

> The worst enemy of life, freedom and the common decencies is
> total anarchy; their second worst enemy is total efficiency.
> —Aldous Huxley, *Tomorrow and Tomorrow and Tomorrow*
> *and Other Essays*[18]

We look back in horror at the colonial days and wonder how we could have been so ruthless. Europeans fanned across the world and claimed vast land masses as their own,[19] and the fact that people might already have been living there was only a minor inconvenience (and in the case of the slave trade, a definite asset). What kind of people were we? What on earth were we thinking? Well, what we were thinking, of course, was that we wanted more resources. We explored the world and found that there were lots of goodies that were undefended, or poorly defended, and we took them. It's not that complicated and it was a predictable step in the development of human societies. Like any other organism, we first consumed what was available locally, and then, when we approached limits to growth, we expanded our territory.

The first step toward the globalization of trade was also made back in the colonial days when Britain wanted to get its hands on the resources of India and China. These countries had resources aplenty, but they also had a rather disconcerting number of locals. Making a profit by direct conquest was beyond the capabilities of even the British military so an innovative little company blazed the trail. The British East India Company gradually monopolized trade in India, increased its influence, and formed its own army. It was not the British army that conquered most of India but the British East India Company's army. Getting into China was even more of a problem, but it was eventually solved with, of all things, narcotics.

British companies bought opium in India and smuggled it into China to trade for Chinese goods. The emperor of China was not impressed; he tried to stop the smuggling, but the British were deeply invested in it and fought back. By the end of the Opium Wars, Britain had bases in Hong Kong and elsewhere to protect its right to trade (or, more accurately, its right to force a country to allow British companies to trade a recreational narcotic). What kind of people were we? What on earth were we thinking?

Well the Indians and Chinese, and Americans (gah), and just about everyone else except the sheep on the Falkland Islands, eventually kicked us out; and the Spanish, Germans, French, Belgians, Portuguese, and Dutch were kicked out of their colonies, too. What to do? The world had filled up and its resources were being more tightly protected. There was no choice now but for countries to trade freely and fairly between one another.

That we call international trade free and fair is, of course, a deceit. Wealth flows freely between rich and poor like water flows freely between a river's mouth and its source.

Globalization has the same goals as colonialism even though the playing field has changed. Now much fuller and subject to much stiffer competition for resources, the world has been shrunk by our sudden abundance of cheap energy. Nonetheless, the rich countries have found another effective way to get their hands on earth's resources. Rich countries now buy their resources from poor countries, but heaven forbid the poor countries should ask to be paid a realistic price. And free trade only applies to the commodities to which the rich countries want it to apply. While poor countries are strong-armed into trading natural resources, intellectual property is protected like the gold at Fort Knox. Oil and timber can be taken from a country, but its people may not follow.

Globalization takes away the "shutdown" triggers of environmental feedbacks. The overconsumption of a local resource ought to cause a slowdown, reducing consumption. In the global economy, however, the resource is found elsewhere and the consumption continues. The overproduction of a pollutant ought to warn a country that it has an environmental problem and cause it to curtail its use of the polluting industry. Instead, we just ship the pollution overseas.

What kind of people are we? What are we thinking?

Globalization is a massive efficiency mechanism that secures resources for the dominant few, but it is also a sign of danger. When productivity and growth from local resources become impossible we begin to source them worldwide, but they are now becoming scarce worldwide. The global economy is stripping resources from all over the world and scattering its wastes into the land, the waters, and the air. A global economy, when it becomes a global machine of efficiency, becomes a global efficiency trap.

Efficient Public Institutions

> We are not a privatization administration, we are an efficiency
> administration, and that's it. It's not theology.
> —Mitch Daniels, former governor of Indiana and
> current president of Purdue University[20]

Public institutions, including governments, should be wary of efficiency. This is something of a hard sell because our disdain for inefficient bureaucracies is almost instinctive. Some people have enormous distaste for institutions, and I'm thinking here of Tea Party–type Republicans and Ron Paul–type Libertarians who argue that government and its institutions are somehow enemies of freedom. They would have us downsize and abolish all manner of institutions; at the very least, they argue for the highest possible levels of efficiency in them, which, in the case of the federal government, would mean limiting its mission to little more than national defense and law enforcement. In truth, the argument is pretty simple (if we ignore the loons), and it's fairly obvious that there can be a fine line between getting the job done efficiently and getting the job done right. We don't want our institutions to be inefficient, but neither do we want them to lose track of their goals. There has been an increasing push for institutions to focus too much on efficiency in ways that weaken us as a society.

As a teacher (of sorts), I think my least favorite word might be *metrics*. Some things are hard to measure, and we do a disservice when we try. This has become a worsening problem in education, where we try to measure all sorts of things in an attempt to make our schools and colleges more efficient. We measure outcome milestones and output milestones (yes, they are different things) and make qualitative assessments of the . . . right, whatever. I once sat on the College of Agriculture Outcomes Based Program Improvement Committee.[21] It took me about three weeks to decipher what the name of the committee meant. There's a place for all this, to be sure, but what's best in education can be whittled away by efficiency. Simple, bullet-point, note-taking teaching with direct, multiple-guess exams is very easy to assess. Free-wheeling teaching with class discussions and essay exams provides a better education, in my view, but causes problems with metrics.

Administrators have always struggled to assess teachers. What constitutes a good teacher compared to a bad teacher? Well, it can be pretty obvious when you're in the classroom, but the folks that assess teachers are rarely seen in the classroom. We make a big fuss about wanting to pay teachers based on performance, but there's no particularly good way of doing it efficiently, so we shouldn't keep making up bogus metrics to pretend that we can. Hire good, well-qualified teachers, give them the support they need, keep class sizes small, and let them at it, I say. Efficient teaching be damned.

And where will efficiency in education lead? Soon you'll be able to get a college degree online without even meeting a professor. Oh, right. I knew that.

My next pet peeve is efficiency in healthcare. We seem to seek efficiency in all the wrong places, ignore obvious needs, and wind up with the most horrendously confusing thicket of bureaucracy. The United States has the best doctors and the most advanced technology in the world, and yet millions struggle to access decent care. Illness can send middle-class Americans into bankruptcy while some of our best doctors are performing cosmetic surgery for the wealthy. The degree to which healthcare has become so unequal is a national disgrace. There are many plot lines in healthcare efficiency, but let me just take up one simple one as an illustration: hospital-acquired, or nosocomial, infections.

We tend to think that people who have a fear of hospitals are just a little wimpy, but they might have a point. There are a number of diseases that you are much more likely to pick up in a hospital than anywhere else. The Centers for Disease Control and Prevention report 1.7 million nosocomial infections contributing to ninety-nine thousand deaths per year in America. It makes sense that hospitals would be a high-risk environment because they host lots of people with infections and compromised immune systems, but that's only part of the problem. The other part of the problem has been that we have created new diseases. To be fair, hospitals have improved a lot in recent years, but we are still finding it hard to control a sudden and unexpected rise in nosocomial infections that began in the 1980s and 1990s.

What has happened is basically a brutal lesson in evolution. Antibiotics started to be used regularly after the Second World War. They revolutionized the control of bacterial infectious diseases. Many of our most important antibiotics were developed very soon after the discovery of the very first one (and still one of the most important), penicillin. Streptomycin was also discovered in the 1940s. The tetracyclines, vancomycin, and polymixin were discovered in the 1950s, and the vast majority of the classes of antibiotics we use today had been discovered by the end of the 1960s. Very few new antibiotics have arrived in the last twenty years. And that's not good, because the bacteria have spent all that time figuring them out. What is disease control to us is natural selection to the bacteria, and evolution knows too well that what doesn't kill you makes you stronger. Some of our plagues of the past are on the rise again.

The easy and efficient disease control offered by antibiotics gave us a false sense of security. We started using antibiotics in lots of situations in which they were not needed, such as to treat viral infections, against which they were useless. We overuse them in surgery to reduce the need for the strictest levels of sterility. Most shocking of all, we use them as growth promoters in livestock. Nearly 80 percent of the antibiotics used in the United States are used to treat perfectly healthy animals.[22] The wonder drugs of the twentieth century are being wasted frivolously, and why? Well, because it's efficient to do so. It's efficient for the doctor to pack you off with an antibiotic just in case your runny nose develops into something worse. It's more efficient for hospital workers to clean hands and work surfaces with an antibiotic wipe than to take the time to scrub them properly with soap and water. It's efficient for farmers to pump their animals with cheap antibiotics so they will produce a little more meat or milk.

Some bacteria deserve special mention. The first is *Staphylococcus aureus*, the nasty bacterium, now also known as MRSA, that is found on the skin of about a fifth of the population. It can cause a range of diseases from very mild skin infections and boils to severe pneumonias, and it is also one of the diseases that can cause necrotizing fasciitis: the flesh-eating bacterium. *Staphylococcus aureus* has always caused these diseases, and it was very difficult to control before antibiotics. Today, more than three-quarters of *S. aureus* infections are caused by MRSA, the methicillin-resistant strain, and we are seeing the evolution of MRSA into VRSA, a vancomysin-resistant strain that will be extremely difficult to control because we are rapidly running out of antibiotics.

Another scary, reemerging disease is tuberculosis. This disease kills millions of people living in poverty, and it's deadly for the immunocompromised (TB is a leading cause of death in people with AIDS), but its rise is also beginning to pose a threat to rich people who can afford antibiotics. There are now strains of *Mycobacterium tuberculosis*, the causal agent, that have evolved resistance to the two mainline drugs, isoniazid and rifampicin.

Gonorrhea is another—and this will get your attention. The CDC announced, in August 2012, that one of our favorite STDs has now evolved resistance to yet another antibiotic. Only one antibiotic still controls it. When resistance evolves to that (and it will: the drug is in the same class as others that have now failed), gonorrhea will be virtually untreatable. Those who contract it will suffer great pain and a range of permanent side effects, and the number of cases will skyrocket.

We'd like to think that the worst-case scenarios will not materialize, and we have great confidence in our public institutions and medical corporations to solve these problems. But that's not likely. The pharmaceutical corporations are hardly even looking for new antibiotics: there's no money in it. Antibiotics are a very poor return on investment because they are used only short-term, until the infection is cured. A much better investment is drugs for erectile dysfunction; there is just as much money invested in making penises erect as there is in curing our most perilous diseases. And our public institutions are compromised as well. Researchers can find easy funding to work with pharmaceutical companies on their profit-motivated research, and the coffers are far too small for research into infectious diseases.

These plagues of the past threaten to become new plagues of the future, and all this could have been avoided, but antibiotics and our healthcare system have become victims of their own efficiency.

It is not the efficiency of public institutions that promotes the well-being of a nation's citizens, but their effectiveness. Indeed, the tempering of the dangerous consequences of unconstrained efficiency is their central role. We have financial regulations to rein in the excessive gambling of lenders and environmental regulations to limit the pollution caused by industry. We have transportation departments to ensure that road and rail networks are built and maintained for the good of the nation and not just for the good of those who can afford to build them, public schools and colleges to educate all the nation's citizens, not just the wealthy, and public healthcare systems to keep the whole nation as healthy and active as possible.

What many people seem to fail to understand about public institutions is their powerful role in supporting the economic competitiveness of a nation. Countries with effective public institutions fare much better in the global marketplace. A better educated nation competes more effectively, especially in science- and technology-based industries. A nation with solid infrastructure, such as roads, produces and distributes goods more competitively. A nation with a strong military can protect its resources and project its power to access resources overseas. The government of a nation, through the creation and maintenance of effective public institutions, is, above all other things, an exercise in global economic competition. In pushing public institutions to be increasingly efficient we tend to undermine their role. Their role is not to be efficient but to mitigate the efficiency traps of unconstrained individuals and businesses.

Private individuals and corporations tend to flourish when they compete effectively, and, as the competition stiffens, they focus on efficiency to stay ahead. They benefit from playing the CCPP game by transferring their costs as widely as possible while holding onto all the profits they can. Meanwhile, they are bolstered by their governments, which design and maintain the public services they enjoy.

The fierce competition of the marketplace is reminiscent of the efficiency-honing attributes of natural selection in the living world. The redistribution of wealth through government services that tempers the market's tendency to oversimplify and lose its resilience is reminiscent of the role of sex. Robust public institutions support us all, and not just the less wealthy. Indeed, robust public institutions are a wellspring of opportunities for business. So, before closing this chapter, we must consider one last wrinkle.

Nations with effective public institutions support competitive businesses and become machines of global consumption. But with resources becoming limiting everywhere, the global economy is becoming a global zero-sum game. What we really need are effective mechanisms for protecting resources internationally, and here our institutions are sorely lacking. The nations of the world, each one honing its efficiency in the global economy, are pushing each other toward a precipice. In the absence of a global effort to temper the efficiency of the global economy it is inevitable that it will grow beyond its means and collapse.

CHAPTER 10
THE ECOLOGY OF COLLAPSE

The rotten tree trunk, until the very moment when the storm-blast breaks it in two has all the appearance of strength that it ever had. The storm-blast whistles thought the branches of the empire even now. Listen . . . and you will hear the creaking.

—Hari Seldon, a character in Isaac Asimov's
1951 sci-fi novel *Foundation*[1]

Our history is a series of periods of societal development punctuated by catastrophic collapses. We don't seem to settle into long periods of stability or to experience gradual contractions. Like the fire-prone forest, we grow, we collapse, and we have to start over.

It is extremely difficult to analyze trends in such a way that we can look into the future and predict what it may hold. We cannot even agree whether the future holds further technology-driven progress or a tough reckoning due to resource scarcities. I'm quite convinced that a debilitating lack of resources will quash technological development and that our societies are headed for failure. But how will the future play out? Are we headed for a long depression or a collapse? When will things start to break down, and what form will the breakdown take?

I very much doubt that most people will understand the reasons for our failure even when it's happening. I see a dangerous future in which our bloated society, finally pushing up against limits to its growth, will fall rather quickly into chaos, but lots of systems will break down and we will have many things to blame other than our own rape of the environment.

Limits to growth and environmental degradation have been at the root of most of our societal failures to date, but we never seem to see them that way. We have been taught that the Roman Empire became consumed by decadence and bureaucratic decay and was conquered by the superior tactics of the Huns, and that the British Empire came to an end when long-enslaved nations demanded and fought for their independence. The Rwandan genocide has generally been reported as the inevitable result of ages-old ethnic hatred. The current turmoil in the Middle East is described as a social awakening demanding democracy and the long-overdue removal of despots. We always seem to downplay the underlying environmental issues.

The Roman Empire was weakened by its own desecration of the natural environment, particularly the depletion of its forests and soils. The Huns could only defeat Rome once it had already crippled itself. The British Empire was weakened by the decline of its coal resources (and the waning superiority of coal in the petroleum interval) and was economically gouged by the Second World War. The genocide in Rwanda was made much more likely when that country became grossly overpopulated, degraded, and impoverished. The Middle East is caught in a collision of environmental and social problems. Failing oil revenues, skyrocketing food prices, unemployment, and overblown populations put a stranglehold on Middle Eastern economies, setting the scene for the removal of despots by force. Bear in mind that Hosni Mubarak (Egypt) and Muammar al-Qaddafi (Libya) were despots for four decades before their people decided it was time for them to go. It just so happens that the al-Assads (Bashar and his father, Hafez), have ruled Syria for four decades.

We are on a collision course. A huge and growing population with massive demands is facing the depletion of many critical natural resources, has reached the global oil peak, and will be increasingly bombarded by the many tragedies of climate change. The relentless growth and prosperity of the last century is coming to an end, but how will it go out—with a bang or with a whimper? Will we prepare and adapt to this new reality, soften the landing, and transition into a new, sustainable future, or will we rage on to the bitter end?

Is the Singularity Near?

> With the increasingly important role of intelligent machines in
> all phases of our lives—military, medical, economic and
> financial, political—it is odd to keep reading articles with
> titles such as Whatever happened to Artificial Intelligence?
> This is a phenomenon that Turing had predicted: that machine
> intelligence would become so pervasive, so comfortable, and so
> well integrated into our information-based economy that
> people would fail even to notice it.
> —Ray Kurzweil, *The Age of Spiritual Machines*[2]

Nearly all our significant technology has been developed since the Industrial
Revolution,[3] and the rate of technological advance has continued to accelerate
to mind-boggling speeds through the petroleum interval. The world has
changed dramatically during the lifetimes of those who currently inhabit it.
This seems normal to us because we have no other personal experience, but
it is not normal at all. My teenage son, Daniel, was shocked the other day to
find out that he was older than Google. How on earth did we find anything
out back in the Dark Ages when his dad was a kid, he wondered. When I told
him that my high school got its first computer when I was in tenth grade he
was aghast. The times they are a-changing.

One measure of the accelerating pace of technological advance has been
in computing speed. Gordon Moore, the cofounder of Intel, observed that the
number of transistors that could be placed on a chip doubled roughly every
two years, and his observation has been dubbed Moore's Law.[4] The idea that
technology improves exponentially seems quite reasonable to many people,
and the consensus seems to be that it will continue to do so. If it does, then we
will quite quickly reach what Ray Kurzweil calls the *singularity*.[5]

Exponential increase is a fairly shocking concept, as exemplified by the
ancient story of Sissa and the chessboard.[6] King Shihram, so the story goes,
offered a reward to Sissa, and Sissa requested a grain of wheat for the first
square on a chessboard and then twice as many grains of wheat for each successive square. The king figured Sissa to be a bit of an idiot until the bags of

wheat started to pile up. By the fourth row, the King could not have fulfilled Sissa's request with all the wheat in India. By the end of the sixty-four squares, Sissa would have owned a pile of wheat about the size of the moon.

Exponential growth is a frightening prospect in just about anything. Exponential growth in living things is soon reined in by the environment, one way or another, but what happens with exponential growth in technology? Is it outside natural constraints? Kurzweil's singularity would have technology soon advancing at a speed that is effectively infinite.

It seems a bit crazy, but one interesting way to look at this is from an evolutionary perspective. Life on earth evolved from the compounds lying around four billion years ago, figured out how to replicate itself, eventually settling on DNA as a code, and combined into complex cells that became large, multicellular organisms, one of which eventually evolved a powerful computer that we call the brain. The brain is perhaps on its way to developing what will eventually evolve to the next level: living machines that can evolve in their turn.

Male bower birds build the craziest structures to attract mates. Some build little roofed huts while others build little avenues of sticks decorated with colored stones, feathers, shells, and the like. It's the coolest thing, but what is really interesting is that the behavior is not learned; it's inherited. This means that the bower bird carries the genetic code that directs it to decorate the environment in a particular way. The shape, structure, and even the color of a part of the nonliving world is inherited through the genetic code of a bird. A similar example is the larval case of the caddis fly, which is camouflaged with fragments of sand, twigs, or gravel. Again, the arrangement of the nonliving materials of the larval case is genetically prescribed by the genome of the fly.

The bowers of the bower bird and the larval cases of the caddis fly are examples of what Richard Dawkins calls the *extended phenotype*,[7] the genotype being the coded genetic information of an organism and the phenotype being the way that organism actually looks, operates, and behaves in the environment. The extended-phenotype phenomenon is far from rare. Termite mounds, anthills, coral reefs, and beaver dams are all examples of an organism's genetic code reaching out of its body to act on the environment.

The reach of the human genome is huge. It's a bit of a stretch, but you can see the human phenotype in a myriad different ways: in our clothes, our buildings, and our machines. Some of our creations are obvious extensions of our own bodies, such as eyeglasses that have evolved into contact lenses and laser eye surgery. Reasonably primitive-looking leg prostheses can be remarkably effective and, attached to a remarkable person like Oscar Pistorius, the South African runner, can extend the reach of the human body seamlessly. How long will it be until it's an advantage to saw your legs off and attach prosthetic ones to go faster? The Cheetah blades that Pistorius uses are surely not the last word. Primitive arm and hand prostheses (picture Captain Hook's hook) are evolving into highly functional computerized ones that link to the nerves of the wearer. Is this the point that we can call the wearer a cyborg, and when will cyborgs play the piano better than the virtuosos of today? And what about the telephone that extended our ability to communicate and then became mobile, a hand-held computer, and is now becoming a pair of computerized eye glasses. Google is working on Google Glasses, and Apple is working on iGlasses. How long before we etch computers directly onto the retina or under the skin or simply hardwire them into the brain?

To talk about "the next step" in technology is already becoming a little irrelevant because many of the next steps are already being taken. The fascinating thing—just as Alan Turing predicted—is now that human and machine bodies, and human and machine intelligences, are actually beginning to merge, we seem to take less notice. Where will this take us? If the advance continues at an exponential rate, it will take us very far, very quickly. Will the machines do a better job of sustaining planet earth? Will they be wiser than us, more compassionate, loving?

I'm not sure what to think of the trajectory of these changes. I do find them disconcerting, I suppose, as well as cool, but I'm not overly concerned about them because I think technology is about to meet a tough reckoning. I don't think technological systems can grow exponentially any more than biological ones can. They are grounded, like all other things, on the energy and resources that they demand. Energy and resources are much more fundamental to progress than is technology, and they are fundamental to the development of technology itself. As Kevin Kelly puts it, "All the mouse clicks in the world can't move atoms in real space without tapping real energy so there are limits to how far the soft will infiltrate the hard."[8]

There is a sort of "race" going on. Will our still-increasing knowledge base and scientific capacity deliver new technologies that can bypass the threats we face from resource scarcity and environmental destruction? Or will we deplete our energy resources and trash the planet? This would arrest our technological development abruptly. Both trends are strong, but which is stronger? Or does it matter? Both are civilization traps.

The Tragedy of the Commons

The tragedy of the commons is Garrett Hardin's explanation of the potentially disastrous consequences of allowing communities to have unregulated access to common pool resources,[9] and it is also the classic exposition of the CCPP game: commonize the costs and privatize the profits.[10] The tragedy of the commons is more than just tragic. It's tragedy in the truest, Shakespearian sense: the remorseless unwinding of a narrative in which disaster is inevitable.

The easiest way to picture the tragedy of the commons is as a shared pasture to which a community of herders has access. Each herder is tempted to increase the size of his flock because, although he may recognize that this will put additional pressure on the grazing land, he calculates that it is in his interest to do so. An additional animal or two will make very little difference, he figures, and he will make more income from adding to his flock. And in any case, the burden of the extra animals will not be his to bear, as the burden will be shared out among all. He is commonizing the costs among all his neighbors and privatizing the profits for himself. It's not an altruistic decision, to be sure, but it is a rational one. The problem, however, is that his neighbors come to the same rational decision and add animals to the pasture, too. Through time, these rational decisions cause the inevitable tragedy to unfold. The pasture is destroyed, bringing ruin to all.

The tragedy of the commons is unfolding remorselessly all around us; in the oceans, forests, oil fields, rivers, and aquifers, and in the atmosphere.

We have concluded that the tragedy of the commons is inevitable wherever communities are left in control of common pool resources. We have tried to solve the problem in two main ways. One method is to impose some sort of central control over the resources by using mechanisms such as quotas to limit fishing catches. The other method is to privatize access to the resource by selling parcels of the commons to individuals with the idea that people will take care of land they own better than land they share: although they will still be able to privatize the profits, they will no longer be able to commonize the costs.

The tragedy of the commons is not always inevitable, however, as has been shown in a number of systems by the late Indiana University professor and Nobel laureate Elinor Ostrom.[11] Professor Ostrom gave a wonderful seminar at Purdue just two months before her death. Despite her frailty—she sat for the presentation—she gave a wise, strong, and engaging presentation, and she was funny and charming, too. "We assumed Hardin was right," said Ostrom, "and that people were trapped in the commons." But through her research and writings,[12] Ostrom has shown that many commons systems have worked extremely well through history, and some work well today.

When we discuss the tragedy of the commons we most often summon up the image of the common grazing lands of a medieval European village (which is effectively what I did above), but Ostrom showed us that this view conceals an important fallacy. The medieval commons actually persisted for centuries and were not ended by the tragedy of the commons. Rather, the medieval commons was destroyed when the European social landscape was thrown into turmoil by the Industrial Revolution. The common grazing lands had never been unmanaged lands exploited by people raising as many animals as they desired. The commons had been managed by comprehensive layers of laws, agreements, and social mores.

Ostrom studied the effectiveness of irrigation systems in Nepal, comparing systems designed and operated by government programs with those built by local farming communities. Various agencies had assumed that the collection and distribution of water would need to be controlled. A number of rather elaborate systems were built, but their concrete channels clogged up and could not be moved to suit changing demands. The villagers' decidedly less flashy mud, log, and brick ditches were much more effective, and they had arranged rather sophisticated social systems to ensure that each member of the community contributed equitably to ditch building and repair and took only his or her fair share of water.

The tragedy of the commons can be a dangerous and pervasive threat, but we also know that common pool resources do not necessarily have to be protected by being separated from communities. The best possible system, in fact, when it can be made (and allowed) to work, is one in which people remain on the land and find ways to manage it as one of its integral components.

Individuals can be rational within boundaries, especially if they understand those boundaries and understand the system feedbacks determining them. Reciprocity is a key norm in successful commons systems, with people learning to trust one another to play their respective parts, but well-defined rules need to be present to control freeloaders. These rules also tend to be highly location specific, and an odd thing that Ostrom discovered was that the rules tended not to actually focus on the particular resource in question but rather on specific issues of utilization. Limiting the use of wood resources, for example, was more likely to be achieved by restrictions on the use of tools, access to certain areas, or the applications for which the wood could be used rather than by delimiting direct quotas of wood.

A rather brilliant example of this kind of approach has been used in lobster fisheries in Maine.[13] One important requirement for the maintenance of lobster populations is to avoid catching pregnant lobsters, but it's hard to avoid the tragedy of the commons when everyone is catching lobsters in private. Fishermen will be tempted not to throw pregnant lobsters back into the ocean unless they are reasonably confident their competitors are doing the same. Enter the lobster punch. Fishermen were encouraged to put a punch hole in the tail of lobsters they caught before returning them to the ocean. The next person to catch the lobster would therefore know that they were not alone in their altruism. Eventually, wholesalers and retailers were convinced not to purchase hole-punched lobsters, and this very simple system was able to prevent the tragedy of the commons.

So there are four possible outcomes when people have access to common pool resources. The tragedy can unfold, ruining everyone; access to resources can be controlled by government; the resource can be partitioned into private ownership; or the community involved can develop an effective management system. The last outcome is clearly the most desirable because it is the only one that maintains both the land and the communities on it. Wherever this outcome can be achieved the potential exists for a sustainable society.

Societies and communities that have achieved this tend to have one very important characteristic: they have had a long time to deal with the inherent limitation of the resource in question. Because the resource has been limited to their community for some time, they have needed to develop a tradition of how to live within its limits. Perhaps it took a former disaster—a former tragedy of the commons—for them to figure it out, or perhaps not, but communities that have figured out the tragedy of the commons breathe rarified ecological air. With respect to that particular resource, at least, they have solved the sustainability puzzle.

Escalation and Addiction

> Any intelligent fool can make things bigger, more complex and more violent. It takes a touch of genius—and a lot of courage to move in the opposite direction.
> —E. F. Schumacher, "Small Is Beautiful"[14]

The tragedy of the commons is rather like an escalating addiction. The consumers of a common pool resource know that their actions represent a long-term threat but they are trapped in consuming it to protect their livelihoods. And as the resource goes into decline, they are trapped into escalating their consumption in order to keep up. Once addicted to something, each minor decision can take you deeper into the miasma. After a while, you look back and see that what once seemed quite reasonable has, step by step, gone sour. A characteristic of drug, tobacco, and alcohol addiction is that what was initially recreational gradually becomes a necessity, much less enjoyable, and harder to back out of.

Even President George W. Bush said it: we are addicted to oil. But it's not just oil: we are addicted to competition and growth, and therefore to all the resources that they require. We can see that this addiction is trashing our planet just like cocaine might trash a life, and we can see that we gain less and less from it. We hate the damage we are doing to ourselves, and yet this is a true addiction. Stopping cold turkey would be too much of a shock to the system and so we continue on. The only thing that will stop us, as is so often the case with an addiction, is when the drug is abruptly taken away.

How will we fare when economic growth is taken from us? It will be tough at first, but we can live, and have lived, just fine without that drug. We will, however, need to be strong through the transition, and we will need supportive communities. Let's start developing that strength and those communities now.

The Overshoot Trap

There appears to be a general consensus that the global population will settle out at around nine billion sometime midcentury. I've heard this statement repeatedly at seminars: it is often the opening statement from which the speaker will then launch into his or her explanation of how the world needs to be changed to be saved,[15] but I have yet to hear a good explanation of why the population should settle out at this particular number. When challenged, people will give some explanation based on fertility rates, or they will explain that the world's population will be universally wealthy by then. I'm not buying it. Overpopulation is one of the great issues of our time, but it is a much more complicated issue than is generally presented.

First of all, we speak of a global population problem, but it is not really a global problem. It is a suite of regional problems. Populations are relatively stable in most of the wealthy nations while they are growing rapidly in most of the developing and poor nations.

Globalization breaks downs walls, promoting the free movement of goods and services, but it strengthens the walls that prevent the movement of people. This is one of the two great ironies (or weapons, if you prefer) of globalization. Globalization enables the "free trade" of wealth and resources from one place to another but prevents people from following. Americans are happy to accept Mexican goods but are getting decidedly antsy about accepting Mexicans.

Further strengthening of the barriers preventing the movement of people should be expected whenever it is in the interests of nations to do so. As we descend into an era of global decline and insecurity, the strengthening of those barriers will become a high priority to ensure that population problems remain regional. The rich countries with manageable populations will increasingly abandon the poor, overpopulated ones. Of much more concern than the global population of seven billion, then, is the Indian population of 1.4 billion, the Chinese population of 1.3 billion, and the Nigerian population of 160 million.

The biggest fallacy about the population problem, however, and a much bigger obstacle to securing a sustainable future, is this oft-cited prediction that populations naturally stabilize when individuals are universally wealthy. It's ridiculous to imagine that the earth could support nine billion wealthy people. We'd need a dozen or more earths to do that,[16] and stabilization is not what happens to rapidly growing populations in biology. Typically, they do not gradually settle out at some balanced number determined by resource limits. They usually overshoot those limits and then crash. Sadly, I see no reason why our human population should be any different.

In 1944, the United States Coast Guard introduced twenty-nine reindeer to remote St. Matthew Island, in the Bering Straits, as an emergency food source. The reindeer population exploded on the small (140 square miles) island. By 1963 it was estimated to be six thousand. The reindeer population then suddenly crashed within a few years and was extirpated. The explanation seems to be that the reindeer starved as their population overshot the capacity of the island to feed them.[17] The carrying capacity for the reindeer was probably less than a quarter of the population that was reached before it crashed.[18]

There is no doubt in my mind that the human population has overshot its carrying capacity in many regions, especially in poorer countries. The idea that our population will just settle out nicely at nine billion has no reasonable basis; it is simply wishful thinking. Much more likely is that billions will die.

One thing that makes the problem of overshoot particularly dangerous is that populations degrade their environments in the process of overshooting carrying capacity. A population of perhaps a thousand reindeer could be supported on St. Matthew Island without causing environmental damage. Part of the reason for the collapse was the slow growth of vegetation in this harsh, northerly environment, particularly lichen, an important winter forage. As the population overshot, the reindeer began to strip the island bare, lowering the carrying capacity in the process. The reindeer population didn't settle back down to a thousand; it failed completely.

Our carrying capacity is falling away from beneath our feet. Our population exploded through the petroleum interval when ever-increasing volumes of fossil fuels raised the carrying capacity of the planet. As we pass peak oil, gas, and coal, that carrying capacity will fall, and when it does, what will we fall back on? Our soils and water? Our forests and fisheries? Like the reindeer gorging on an apparent abundance of lichen, we will find that we have not only run off the edge of the cliff, but that the ground is falling away at the same time.

Burning Bridges

Napoleon's unconquerable Grande Armée marched into Russia in 1812 with a six-hundred-thousand-strong invasion force. By the end of its disastrous retreat through the Russian winter, more than 90 percent of the soldiers were dead. The Russians carried out a brilliant strategy. They avoided as much direct confrontation as possible and drew the French deep into the country. The French won the famous Battle of Borodino outside Moscow, but the Russian army withdrew and scattered before it was destroyed, and then the Russians burned Moscow. The French were starved of supplies deep inside Russia and had no choice but to retreat, but they had not planned for this. Supply lines were inadequate and the Russians skirmished with them repeatedly, slowed them down, and allowed the harsh Russian winter to do the rest. The most powerful army in the world was annihilated and the French Empire was doomed.

The problem that Napoleon encountered is all too familiar. It can be much harder to retreat than advance. It's easy to climb up that rocky mountain outcrop than it is to crawl back down again when you get lost in the fog. It's much easier to get into a relationship than it is to get the heck out when it starts to go sour (or is that just me?). None of this should be particularly surprising. It's in our nature and, in fact, in the nature of things.

The appendix, that more-or-less useless appendage in the human gut, prone to inflammation and explosion, is often used by creationists as a disproof of evolution. If evolution is supposed to constantly hone organisms toward some state of biological perfection, it is parroted, how on earth does one explain the appendix? (It's a nutty argument, of course, because you could also ask why an omnipotent God would saddle us with the darn thing in the first place—but never mind that.) Well, the appendix is quite easily explained as an evolutionary hangover from the eons our distant ancestors spent as herbivores. The appendix evolved to digest cellulose, and if animals like the rabbit are anything to go by, it probably did a pretty decent job for us. In our more recent evolutionary history, however, we have become increasingly omnivorous and have had less need to digest cellulose. Even the vegetarian part of our diet has come more from easily digested foods such as grains and fruits rather than leaves.

If we could have a mulligan on the design of the human body we would rework ourselves without the appendix, and, let's face it, a bunch of other things. I'd appreciate a redesign of the spine to limit my back pain. (My brain could use some rewiring, too.) But there are no mulligans in evolution. Evolution is a process that operates generation by generation. It does not have the opportunity to reflect periodically and perform a systems overhaul. It acts only on the transition from the current generation to the next. Nature can seem perfect, but you don't have to look all that hard to see that it is not. It is a constant struggle that operates baby step by baby step, and it often leads to unfortunate dead ends. There are far more species that have played the game of life and lost than there are still playing. Evolution repeatedly burns the bridges over which it has taken species, and they can never go back.

A city is not much different. Consider London as an example. Settlements have existed on the River Thames in the vicinity of London for thousands of years. Archeological excavations date the remains of a bridge or jetty in London's Vauxhall district at three thousand years old. The Thames plain has some of the most fertile soils in Britain (shame, then, to have concreted over them so completely), and it should be no surprise that people would settle in this fertile landscape. The Romans established Londinium at this site directly after their invasion in 43 CE and built a bridge across the Thames. A bridge still stands at the same site: London Bridge.

Numerous upgrades have been made to ancient Londinium, but, as with biological evolution, the evolution of this rural backwater into a global hub of more than sixteen million occurred in baby steps. Had some Norman bureaucrat around the time of the Magna Carta been foresighted enough to recognize that the population of London would increase by three orders of magnitude it might have been possible to plan for the future; or perhaps the government of Tory prime minister William Pitt the Younger could have tackled some long-range planning two centuries ago, when the city's population was reaching its first million, but neither of these administrations laid down a particularly effective long-term plan for London. London has simply grown and evolved, layer by layer, and it cannot be overhauled. If you want to travel from the ancient kingdom of Wessex to the ancient Danish territory of Guthrum today, you may well cross the Thames where it has been crossed for two thousand years: London Bridge. But watch out for the traffic.

The center of London is more or less where it has always been because that's where the bridge was; so that's where the roads appeared and where the businesses developed. When it got busier another bridge went up nearby, and more roads, so more businesses arrived. Then came cars, more roads, more businesses, more roads, and here we are, stuck in traffic, wondering what idiot designed this place.

So here's a question: Why are train tracks the width they are? It turns out that it's not because this is necessarily the best width; rather, it's because the first tracks were built to fit old carriages that were built to fit old roads that had been built to fit the carriages that had come before, which had been built to fit old Roman roads that had been built for chariots that had been designed to be pulled by horses. Yes, the modern train track was designed thousands of years ago by a horse's ass.[19]

Towns and cities can be designed, and have been, and sometimes the designs are effective, but the dominant force has not been one of development but of evolution, and evolution never looks back, never analyzes, and never plans for the future.

Apply this logic, now, not just to the evolving architecture of a city but to the myriad interrelated, evolving systems that make up our complex modern world. There are a million London Bridges that have found themselves mired in, and yet essential to, a cumbersome system that cannot be undone. Technological advances bring progress, but they are usually built on the remains of what came before: it is generally more efficient to do so. We become increasingly dependent on the new system, which must get bigger and better, but there is no easy way to go back. If we were to need the Thames Plain to produce food, for whatever reason, that would be tough luck. It's covered in concrete. We become increasingly dependent upon the innovations and systems that control our lives, and each step forward burns another metaphorical bridge over which we might wish to retreat in the future.

Many civilizations have collapsed, and they all burned bridges. The Hisatsinom of the American Southwest built an extensive society but degraded their environment so badly that there was virtually nothing left to even hunt and gather when the end came. They abandoned the region, leaving the Dine, who arrived in the region centuries later, to wonder who had left mysterious ruins in the desert.[20] Polynesians canoed across thousands of miles of open ocean to find a forested Easter Island, but when the forests were depleted, bringing about their civilization's collapse, they could no longer build canoes; they were left trapped in the middle of the Pacific Ocean. The Romans devastated the Mediterranean region so completely during their experiment in civilization that five centuries of Dark Ages followed.

Our prodigious efforts in bridge burning are making the Hisatsinom, Easter Islanders, and Romans look like part-timers. Forests, soil quality, water resources, farm land, water quality, air quality: we are damaging them all beyond repair. We are paying lip service to their conservation as we pretend to protect them with efficiency measures that degrade them still more. What will the world look like by the time we have finished this latest experiment in civilization?

Efficient Control, Efficient Collapse

> Surely our understanding of the nature of man or of the range
> of viable social forms is so rudimentary that any far-reaching
> doctrine must be treated with great skepticism, just as skepticism
> is in order when we hear that "human nature" or "the demands
> of efficiency" or "the complexity of modern life" requires this or
> that form of oppression or autocratic rule.
>
> —Noam Chomsky, *Notes on Anarchism*[21]

Societies collapse when a bloated society depletes needed resources. Demand
grows and supplies shrink at the same time. We scratch our heads and wonder
why on earth they didn't see the problem coming—but perhaps they did.
But if that's the case, we must scratch our heads twice and wonder why they
didn't do anything about it. They did do something about it, of course:
they tried harder and harder to keep the system running, pulled out all the
efficiency stops, and plowed on. And then, as the end approached, they fought,
desperately, to quell their own populations and repel usurpers.

The Easter Islanders put themselves out of business by systematically
stripping their little, isolated island of resources, but the magnificent Moai,
those iconic, giant, ten-ton stone statues, were still being carved. Dozens of
them lie unfinished in the Rano Raraku quarry to this day. The rulers of Easter
Island were still trying to demonstrate their power to the very end by building
the biggest and best Moai, even though the last of the trees that might have
been able to transport them were being felled.[22] Finally, the Easter Islanders
gathered into two armies that fought it out in a final battle on the slopes of the
dormant volcano, Poike.[23] The winners didn't win much.

The Romans built one of the biggest empires and had one of the most powerful armies the world has ever seen, but they consistently overdrew from the
ecological savings bank. As they did so, they had to repeatedly expand the empire
to capture more resources. The degradation grew worse and the area to defend
grew larger. The empire began to struggle. They controlled their own people
with all manner of methods, one of which was to put on lavish gladiatorial events
at the coliseums. They controlled the periphery of the empire with brutal force.

Displays of power were everywhere, but their real power was gone. Attila the Hun, Alaric the Visigoth, and the other "barbarians" eventually picked them off.

Perhaps the most interesting displays of precollapse power are those of the Maya. This was a civilization that filled Mesoamerica with great cities from the Mexican Highlands to the tip of the Yucatan Peninsula and northern Honduras. They built extensive canals, dams, and observatories, invented the famous Mayan count calendar, and developed a sophisticated style of writing. At the peak of their powers they began to erect magnificent pyramids, carved stelae, and, at Copan, began the magnificent hieroglyphic stairway. The biggest pyramids and most ornate carvings, however, were the last things they ever built. Conflict escalated among the Mayan cities, the latest carvings became increasingly patriotic, and then the Mayan civilization collapsed almost simultaneously across the entire region.

They put missiles on rooftops to protect the Olympic Games in London, 2012.

They insist that I turn off my laptop when a plane is taking off and landing. Why is that? Can it really make the plane crash? I don't believe it: I'm forever leaving my cell phone in the overhead bin and forgetting to turn it off, and we haven't crashed yet.

The United States has nearly 1.5 million active-duty military personnel in more than 150 countries.[24]

On a recent trip to China I tried to count the surveillance cameras in Tiananmen Square but there were simply too many. In Lhasa, Tibet, it was easy to count the number of surveillance cameras on our tour bus (one) and the number of places in the city that I was hidden from video surveillance (none). I got some great photos when soldiers weren't blocking the view.

It's against the law in some states to suggest that the US food supply is unsafe. It's called "agricultural disparagement." Now who would want to disparage our perfect agricultural industry? Not me, that's for sure. I'm happy that the pink slime is keeping me safe, yes sir.[25]

Roughly 3 percent of adult Americans are currently in jail or on probation or parole. The figure is nearly 10 percent for African Americans.[26] And why do armed bank robbers often get longer jail sentences than child molesters and rapists? Successful bank robbers, by the way (and there are precious few of them), make an average haul of around four or five thousand dollars, and injury is caused in less than a twentieth of robberies.[27]

Big brother is definitely watching you as you stroll around Tiananmen Square in Beijing, China. *Courtesy of Steve Hallett.*

One of my favorite books is Naomi Klein's *Shock Doctrine*,[28] which explains a disturbing phenomenon that I had not noticed until I read it. Once you see this, however, you'll see it everywhere—and be disturbed. Klein shows how governments use political shocks such as terrorist attacks and natural disasters as an excuse for pushing through legislation that takes away civil liberties and bolsters the power of the already powerful. The aftermath of the Pacific tsunami in 2004 saw land "redeveloped" by tourist resorts. During the Iraq War, a new Iraqi law promoted the rights of foreign oil companies. A population shocked by a civil or natural disaster is ripe for the imposition of what Klein calls "disaster capitalism." The granddaddies of them all, of course, were the unrelated invasion of Iraq, the stripping of habeas corpus rights from captives taken in the "war on terror," the use of torture by American interrogators, and the advent of warrantless wiretapping, all of which followed the shock of the World Trade Center attacks of September 11, 2001. The population is disoriented: secure the oil and impose tighter civil control.

Security cameras in Tiananmen Square might not be such a surprise, remembering the 1989 riots, but why is there a security camera here? This is a nunnery in Lhasa, Tibet. *Courtesy of Steve Hallett.*

The world is teetering on the brink of disaster, the people with their grip on economic power are pushing us faster toward the edge of the cliff, and we are all being swept along. We should all be enraged, and some of us are, but many are not. Why is that? And it's even weirder than that: many of the people who should be the most enraged seem to have decided it's all a good idea. A combination of biased, monopolistic media conglomerates, like Rupert Murdoch's phone-tapping bully News Corporation, and manipulative politicians that have somehow conflated conservative economics with conservative so-called moral values seem to have duped millions. Tom Frank calls people's eagerness to be manipulated in this way a derangement: "This derangement is the signature expression of the Great Backlash . . . that mobilizes voters with explosive social issues—summoning public outrage over everything from busing to un-Christian art—which it then marries to pro-business economic policies. Cultural anger is marshaled to achieve economic ends."[29] Or perhaps Stephen Biko said it best: "The most potent weapon in the hands of the oppressor is the mind of the oppressed."[30]

The people who like things the way they are have total control over the system so that they can keep the machine churning forward. We are not going to change the world. We are not going to change direction. All you can do is protect some small part of it and begin to prepare for a very difficult future.

The Ecology of War

> I do not know with what weapons World War III will be
> fought, but World War IV will be fought with sticks and
> stones.
> —Albert Einstein, 1948, in a letter to President Truman[31]

Wars are caused by foolish or power-hungry leaders fighting over ridiculous ideologies, by historic enmities, and by conflicting social norms, and they break out suddenly and cannot be predicted: or at least that is what we have been told. Perhaps war can be understood from an ecological perspective. War is an example of what an ecologist might call a "stochastic event": an anomaly in a complex system. Ecologists may not be able to predict a specific stochastic event, but they can describe the conditions that make them more likely. War comes to a society like fire to the forest, and it is at its most likely when the political and economic deadwood has accumulated for too long beneath the efficient, controlling canopy of the dominant few.

When I stopped to take a photo of a nuclear power plant in Illinois this police officer (calling in my ID) was there in less than a minute—I'd say less than thirty seconds. He turned out to be a real nice guy; we had a fun chat. Although he didn't say so, I got the distinct impression that he was less than impressed that he (a public servant) was at the beck and call of Exelon (a private corporation) to check out anyone who stopped nearby (on a public road). He implied that I shouldn't take photos of the plant, but when I asked if there was any law that prevented me from doing so he said there wasn't.
Courtesy of Steve Hallett.

In the 1940s, a series of tense, sometimes hostile negotiations about the price of oil began among producers. Negotiations developed into skirmishes as producers found themselves at odds with consumers: Operation Desert Storm, in 1991, was the first significant oil skirmish. When global oil production stagnates and then goes into decline, the game of global oil politics will change again. Powerful consumers will compete for declining oil supplies in volatile parts of the world. Will oil skirmishes become oil wars?

If peak oil were the only major challenge this century, things might not be too scary: but it is not. We also face soaring population pressures, declining freshwater resources, and failing fisheries—to name a few—and each of these could also ignite local and regional conflicts. Meanwhile, the planet will continue to warm and the impacts of global climate change will follow the earthquake of peak oil like a tsunami. The impacts will be meted out unevenly. Some regions will escape more or less Scot-free while others will suffer unbearable heat stress and crushing droughts.

The twenty-first century will be one of enormous change, with roiling problems set on a collision course. These are difficult and dangerous times. That our civilization is reaching its limits to growth can no longer be doubted, and it seems inevitable that declining energy and natural resources will not only limit growth but will set the stage for a significant contraction. But what form will that contraction take, and on what timeline will it occur? Maybe the downturn, when it comes, will be long, drawn out, and steady. It is also possible, of course, that it will be sudden and punctuated with conflict.

Meanwhile, we continue along the road we know. The idea that this latest experiment in civilization may be just about played out is still the minority view. Most people still seem to think that the problems we face will be solved and that technology will save the day. It seems unlikely that any sudden technological "liberation" will come because that is not what we need. We are not reaching the limits of resourcefulness and technological brilliance; we are approaching limits to resources. Indeed, we have overshot them in many ways already. Our population will not settle out at some mysterious number that ensures stability and wealth across the globe. Frankly, we already need more earths to achieve that. We have burned many bridges in our insatiable quest for growth. Our quest to control our world, and to keep it producing, has blinded us to the perils we now face.

I do a thing called "tragedy week" in an honors class I teach at Purdue University. I divide the class into four groups (about six students per group), each group representing a village and each student representing a herder. They are the only inhabitants of a grassy island in the middle of the ocean. I give everybody fifty sheep at the start of the exercise and tell them that they can either add sheep to their flock or cull their flock every spring.[32] The objective of the game is to support their families as best they can by raising sheep. I first assumed this game would demonstrate the tragedy of the commons in a single, forty-five-minute class, but weird things started happening, which is why the game has become "tragedy week."

I've done this exercise four times now and the student herders and villagers have done the same thing every time: exactly what is predicted by the tragedy of the commons. The herders begin to add to their flocks in order to increase their wealth. After a couple of years of increasing their flocks I show them that the quality of the pasture of the island is beginning to decline.

There are a number of solutions to this problem. The herders can build fences to enforce the feedbacks of overgrazing directly on the overgrazers, they can create a central government that enforces strict sheep quotas, or they can devise a set of sheep-management rules that everyone obeys (or is forced to obey). The students choose none of the progressive options. They add more sheep and the tragedy continues unabated.[33] Then, around year five or six, as the situation becomes perilous, the oddest thing happens. It has happened every time I've taught the class (and without my suggestion). One of the villages "invades" another village and either kills everyone, kills or steals their sheep, or both.

This causes great hilarity, of course, and a quarter of the class now has to watch from the sidelines. I let the slaughter proceed and then let the game continue. The tragedy, of course, is solved only temporarily. Sheep numbers climb again, the pasture begins to degrade again, and, as often as not, war breaks out among the three remaining villages. It's a teachable moment, and I love teaching this class because we can now settle down for a while, think about what just happened, and see if there might have been any, er, better solutions. The herders finally figure out that there are inherent limits to growth and find ways to manage their communities on the commons.

It's fairly shocking how easily we choose the path of conflict even when much better options are eminently feasible. We shouldn't be shocked, of course: it's what we usually do.

PART 4

BEYOND EFFICIENCY

What Will You Do in the Thirties?

My view that a global crisis is coming seems to have made me a bit depressing to talk to. (I know: some people have no sense of humor.) I got into the habit of talking about the "End Times," partly as an ironic sleight on the notion of a biblical end times. Anyway, my love, Shelley, grumbled at me one day, "Enough with the 'end times' thing, honey: it's a bit depressing. Can you at least find a different name for it?" Sure I could. "How about 'the Thirties'?" I suggested. That sounded much better, she said, until she pondered it for a few seconds and asked why "the Thirties"? That's my best estimate for when the global economy will crash. "Thanks, honey, that's so much less depressing. What's for supper?"

It's impossible to predict what's going to happen over the next few decades. We might continue to stagnate and contract, we might get our economies growing again for a while and then crash dramatically, or the crash may involve smaller or larger conflicts of any number of kinds, including nuclear exchanges. We might also figure things out and save the world as it is, I suppose, but I'm not betting on that. Planning your life requires at least some kind of overall timeline for the future, but our visions and timelines range from one extreme to the next. Here's mine . . .

All trends are moving in the wrong direction: a collision course. And as for this experiment in civilization: stick a fork in it, it's done. Population and resource demands are still in runaway mode while resource supplies are failing. The environmental damage caused by our desperate quest for resources continues to escalate. Global climate change is no longer controllable, and its impacts, which are already beginning to bite, will get worse.

213

The most indicative trend is energy resources, and I think the failure of oil, natural gas, and coal production will be the factor that tips us over the edge and triggers failure. I'm predicting that oil will be in obvious decline by the end of this decade. Natural gas will fill some of the void until it also peaks toward the end of the 2020s. Coal will still be a major energy source, but in fewer countries. By the 2030s the entire global energy supply will have gone into serious decline. This will break the economic dam that allows the pent-up energy of all our other problems to flood in.

I think the global economy has already peaked and will never rise back above the economic heights of the last quarter century. We will stay on this bumpy plateau for a few more years, perhaps another decade, and then serious economic failures will come. China and India are particularly vulnerable, and I think economic failure is coming to both those countries soon. By the end of the 2020s we will be hanging on by the skin of our economic teeth. Then, in the 2030s, all hell will break loose. Whether or not I'm right remains to be seen, and I'll certainly change my benchmarks as the years go by and the data come in (a nuclear war next week would change my view of the world), but this is the timeline I am working from.

So what will you do in the thirties, and what is your plan for getting there?

I hope this book will help you to fix your own benchmarks for surviving the thirties, and perhaps even thriving. Much more important, however, will be to put yourself in a position to respond to a wide range of eventualities because no matter what predictions you make, many of them are likely to be wrong. Resilience is the key. Here are a few things to consider:

- Debt. If you are in debt this might be a good time to get serious about fixing that.
- Housing. Will you be able to stay in your house or apartment? How well will your home function if there are blackouts? Will you be able to stay warm in the winter and cool enough in the summer?
- Employment. Will your line of work be considered necessary during a deep economic failure? (Hedge fund managers beware; trades people rejoice.)

- Food. Where will your food come from? Will you have access to what you need and will you be able to afford it? Will you need to grow some of your own food? Where and how will you do that?
- Water. Will you have access to the water you need?
- Security. How secure is your community, and how secure will it be in a crisis?
- Transportation. How much will you need to move around? A long daily commute may become a huge problem. Switching to a more efficient car is not as big a priority as it might seem. Keep the car for now and worry, instead, about what you'll do after the roads are too potholed to drive it.

Some of the things that we have been fretting about for years might not be as important as they seem. Stop trying to save energy for the sake of the world's energy supply, but worry, instead, about your own energy needs. Also, consider your community's and your city's access to energy in the future. Can you reduce your usage? Will you need more capacity? More efficient appliances are a good idea, but focus on how they will enable you to function in the thirties rather than on just their inherent efficiency. Worry less about your carbon-dioxide emissions and more about the impacts climate change will have in your region.

If you're a wealthy carpenter living in a simple stone cottage that is paid off, with two hundred acres of high-quality land in a nice community of people you know and trust, near a small, thriving town, you might be in pretty good shape already. Perhaps you also recently installed geothermal heating, solar panels, and a small wind turbine between the greenhouses and the chicken coops. You still have problems, not the least of which is that I'll be coming to live with you in the thirties, but you are in good shape. On the other hand, you may be scraping by in a sketchy suburb of a big city, commuting three hours a day to a desk job at the Ferrari sales room, and wallowing in debt.

The time has come to take a serious look at your situation and how you can improve it. Part of the difficulty you face may be that the "to do" list is rather long, but there is still time, I hope. A particularly hard decision in preparing for the thirties might be your location and employment. Lots of us live in rather expensive cities and suburbs because of our work, but I think you should look very seriously at the city in which you live and think about what it will look like during a collapse. If it looks like trouble, start planning to get out.

It's often very difficult to make such changes, of course, and it may not be possible to relocate. In that case, you must do what you can to increase the resilience of your own personal bubble, but it may be even more important to consider the level of organization that surrounds you. We will focus more on this in the last chapter, but your community could be vital. Developing strong, diverse, cohesive communities will increase your personal security considerably and will increase the chances that the services you need will still be available.

CHAPTER 11

RESILIENCE: BEYOND SUSTAINABILITY

With the coinage of "sustainable development" the defenders of the unsteady state have won a few more years' moratorium from the painful process of thinking"

—Garrett Hardin, *Living within Limits:*
Ecology, Economics, and Population Taboos[1]

The term *sustainability* has been overused in recent years, and its value has been diminished as a result. Sustainability is the ability to be sustained; to remain intact indefinitely, but the term has been applied so broadly and carelessly that it now seems acceptable to call almost anything sustainable so long as it appears to deliver some kind of environmental improvement. A building that uses a proportion of its energy from renewable sources or has been retrofitted to higher insulation standards is likely to be called a sustainable building. A farm that uses reduced volumes of synthetic fertilizer or pesticides is likely to be called a sustainable farm. Any number of changes could be counted as sustainability improvements even though they do not necessarily ensure sustainability in the strict sense: the long-term survival of the system.

This problem with the term is fairly obvious, especially in advertising. A quick search on Amazon.com will find you (in addition to books by Steve Hallett) sustainable jewelry, a sustainable "slippery when wet" sign, sustainable eye shadow, sustainable skateboard wheels, and sustainable pet toys. I also searched "sustainable sex toys" (just for research, you understand) and came across a few interesting things (the phthalate-free phallus had me giggling for hours). Stefanie Weiss's *Eco-Sex: Go Green beneath the Sheets and Make Your Love Life Sustainable*, looks like a must-read. I don't mean to be disparaging about these products. I imagine they are probably better for the environment than the "unsustainable" ones, but they demonstrate how the term has been diluted.

But there is a much more important question about sustainability (real sustainability, that is): Is sustainability actually what we want? What are we trying to sustain? Are we trying to sustain the world as it is? There are certainly things that we would like to sustain, such as our soils and our freshwater supplies, but sustain this modern world? I don't think that's a desirable goal because it implies that we should try to sustain this modern throughput economy. No system that consumes the volumes of resources that we consume and dumps the volumes of waste that we dump can ever be remotely sustainable. Attempting to sustain this system represents a vain attempt to keep in a steady state something that is not even close to being in equilibrium. We cannot sustain this model of civilization and we should not even try. The longer we succeed in keeping it alive, the more damage it will cause and the more dramatic its eventual release will be.

This, then, is our most pervasive efficiency trap. By promoting efficiency as a means of keeping this system alive we merely deepen the environmental trauma that we are causing. If failure—release, collapse—of the system is inevitable at this late stage of its maturation, resilience is much more important than sustainability. Sustainability might have paid dividends half a century ago, before we reached this stage of brittleness, and it could pay dividends in the future, as we attempt to rebuild, but the emphasis today should be not on preventing collapse but on preparing for it.

Biomimicry

The elegance with which evolution has solved riddles in the natural world, such as locomotion, flight, vision, and data processing, has stimulated many inventors to attempt to mimic nature to produce new innovations. Leonardo da Vinci produced detailed drawings of theoretical flying machines that were based on deconstructions of the mechanisms of the flight of animals.[2] The inspiration for Velcro was the barbed hooks of burrs, such as the burdock and cocklebur, whose fruits so eagerly grip clothing and animal fur. Engineers are trying to re-create photosynthesis in human-made devices, biotechnologists are trying to re-create the remarkable pharmaceuticals that the plant world already figured out, and computer scientists are eager to develop processors that can compete with the human brain. The natural world has evolved untold biological wonders that can be mimicked. The most important benefit from mimicking nature, though, might not be an advanced technology that mimics a particular biological trait, but social and economic systems that can mimic the deep-seated robustness and resilience of ecosystems.

Economic systems have a lot in common with ecological systems and we can learn a lot about their functioning when we analyze them as such. The development of businesses and societies resembles an ecological succession. The diverse cadres of entrepreneurial start-ups that proliferate when a new technology comes on the scene tend to merge into a few, large, cumbersome, and dominant companies in the way that the colorful gaggle of weeds and wildflowers is eventually overtopped by a few dominant tree species. The failure of societies and the collapse of civilizations resemble the release phase of a cyclic succession such as a fire-prone forest when it finally bursts into flames.

At its core, an ecosystem is a web of energy and resource flows, and the apparent complexity of an ecosystem can actually be boiled down to a few simple processes. Where does the energy that drives the system come from, and how is it dispersed through the system? Where do the resources needed to construct the system come from, and how are they dispersed? Which species capture the energy and resources, how much do they capture, and when?

All ecosystems are powered by the sun, and solar energy is captured by plants (and algae and cyanobacteria).[3] All the energy of an ecosystem enters by the conversion of solar energy to chemical energy in photosynthesis. The energy flows into the rest of the ecosystem when plants are consumed and more organisms consume and parasitize one another. The energy of the sun drives the whole ecosystem and is eventually lost as heat. An ecosystem, then, is an energy-throughput system that depends on the sun to keep shining consistently. An extremely important characteristic of ecosystems is that the amount of energy they receive is the same each year and each millennium.

Human systems are also energy-throughput systems, as are all systems, as per the laws of thermodynamics. Our primary energy sources are captured and fed through our systems, making them go, and here is where we encounter the first troubling difference between human systems and ecosystems. While the amount of energy flowing through an ecosystem is the same every year, the amount of energy flowing through human systems can change dramatically.

Ecosystems are strictly bounded and have been challenged to survive within the same, inalienable energy limits for eons. Operating at the limits to growth has forced ecosystems to solve the sustainability puzzle, and they have done so in a range of different ways. Human systems must also operate within inalienable limits, but we have been seduced into believing that those limits will always increase because over the last few centuries they have. We have constantly expanded our systems toward the limits to growth (as ecosystems would) and have become convinced that we will never exceed them. But our energy throughput is about to meet a tough reckoning. When our limits cease to expand and instead begin to contract, we will face a sustainability puzzle that no system could solve.

Many of the other resources of ecosystems operate in a completely different way from energy. The nutrients from which organisms are constructed are recycled. We are all built from whatever chemistry happened to be lying around since life evolved on planet earth.[4] What is waste to one organism is food to another, and all living things are trying to get their hands (pseudopods, tentacles, roots, and leaves) on carbon, nitrogen, phosphorous, potassium, sulfur, and all the other elements from which they are made. They just have different ways of doing it. The plant specializes in the collection of carbon dioxide from the air and microbially prepackaged nutrients from the soil, deer eat the plants, mountain lions eat the deer, dung beetles and a suite of microbes consume the plant, the deer, and the lion when they die. Their dead bodies and poop are prepackaged back into a pool of food in the soil that goes back to the plants.

Human systems are not recycled in this way. Ours are throughput systems for resources as well as energy, and this is very bad. We have a long way to go if we want to mimic the recycling of a forest or a prairie, but all systems are bounded by the limits of needed resources. We seem to think we are an exception to this rule, but we are not.

Ecosystems are deeply sustainable. An energy supply is provided with absolute reliability and all other materials are recycled. Ecosystems are strictly bounded. The amount of energy coming into the system is fixed and the other resources are available within very tight limits. Ecosystems are, as a result, pervasively competitive, and all species must strategize and battle to get their share. The diversity of these strategies and the intensity of competition appear chaotic, but despite the chaos within, ecosystems have very simple rules and strict boundaries, and they remain surprisingly stable. Understanding which species will acquire the energy and resources as well as the extent and timing of their energy and resource acquisition, is complicated, but the system as a whole collects, distributes, and recycles in a very predictable way.

The reason that a fire-prone forest repeatedly goes through its cyclic succession of development, maturation, release, and recovery is easily understood within this simple, bounded framework. It reaches a point, at a late stage of maturity, when a few, large species have efficiently captured the lion's share of the energy and resources in their combustible bodies. Fire releases the energy and nutrients that have been locked up, and the cycle starts over. But not all systems go through wide-scale release phases. Some ecosystems remain in a stable state of maturity for very long periods of time without experiencing widespread, periodic booms and busts.

One such ecosystem is a rainforest. What happens here? Well, the rainforest is still a simple system powered by a steady input of sunlight and built from recycled resources. The rainforest is rarely razed by a major disturbance such as a fire, and rainforests spend a much longer time period in a state of maturity. Disturbances come more often but in smaller patches.[5] If an old tree is toppled by the wind, the familiar process of succession will swing into action. The weedy, r-selected pants rush in, the shrubs follow, and then the K-selected trees finally restore the canopy gap. The system remains bounded, of course, and so the battle for energy and resources rages on. Each species is constantly tested by natural selection, and its strategies and skills are constantly honed. The result is the most diverse set of ecosystems in the world. It appears to be the most chaotic, complex, and confusing of all, but it is just as simple as any other. It collects and distributes energy. It recycles resources. If we want to emulate any system, it would be this one. Its diversity and beauty constantly grow, and there is no lack of opportunity for innovation and competition, yet it remains in a steady state. How do we build a civilization that resembles a rainforest rather than a fire-prone forest? It's probably not possible in the long term without a steady supply of energy and sustainable levels of resource use, but figuring it out for future generations is a worthy goal.

The diverse, multilayered canopy of the beautiful rainforest that cloaks the hills of Tobago in the Caribbean. *Courtesy of Steve Hallett.*

The fire-prone forest and the rainforest are both highly resilient. They go on and on. They just have different ways of doing it. The rainforest appears to be more resilient than the fire-prone forest because it lacks such a dramatic release phase, but the apparent lack of resilience in the fire-prone forest is just an illusion. The model of the fire-prone forest is much more dramatic, however, during its transition from maturation to release, and it is this part of the cycle that our civilization now faces. We don't face extinction, but we do face collapse and a long time in the weeds.

A Steady-State Economy

> Questioning growth is deemed to be the act of lunatics,
> idealists and revolutionaries. But question it we must. The
> idea of a non-growing economy may be an anathema to an
> economist. But the idea of a continually-growing economy
> is an anathema to an ecologist.
>
> —Tim Jackson, *Prosperity without Growth:*
> *Economics for a Finite Planet*[6]

It has been the fate of human societies to cycle like the fire-prone forest rather than flourish like the rainforest, and we find ourselves at the most dramatic and unpredictable phase of the cycle. We have accumulated resources in the hands of the wealthy few and depleted them from the rest of the system. We are replete with accumulated deadwood. The fire is ready to burn, and, once it is lit, it will become a wildfire.

Early economists were aware that wealth was rooted in natural resources, or at least in land.[7] The accumulation of land and the natural resources that it held (and the colonization of regions that possessed it) was a sure path to wealth. The central role of land was displaced by a focus on labor and capital as the origins of wealth as we approached the twentieth century.[8] This is understandable when we remember that resources, particularly energy resources, had never reached practical limits. The emergence of neoclassical economics took the discipline a step further from reality and stripped its connection to natural resources almost completely. It focused on the exchange of wealth rather than on its creation. That there would always be an abundance of wealth was assumed.[9] Whenever it was necessary to include some function of the environment into their calculations, the neoclassical economists tended to consider needed resources to be unlimited or infinitely substitutable. Environmental damage, meanwhile, was an afterthought; an *externality*.

Economists have been completely blindsided by the fundamental throughput nature of the systems they analyze. They have failed to acknowledge that for wealth to be exchanged, it must first be created, and that resources must be drawn upon to make that happen. Our economic growth has been predicated upon a throughput of resources that has always increased, both in volume and efficiency. We have always had more resources, and we have always improved the efficiency with which we exploit them. Economists have gradually forgotten that ecology is the foundation of economics.

Neoclassical economics has been working from the wrong baseline. It has operated with the rapid growth of the last two centuries as its "constant," but the baseline has not been constant at all. It has been an upward curve. Neoclassical economists have been fooled into thinking that this trend will continue, and so they extrapolate into a future using this same baseline. Their predictions, therefore, are deeply flawed. The upward curve has been driven not by the exchange of wealth or even by the availability of labor and capital, but by access to resources. A new economics that operates on a flat baseline— a steady-state economy—has been needed for decades, and an economics that accepts declining resource availability is needed to replace even that.

A number of people have discussed the need for the development of a new economics that can support a steady-state economy. Two of the most insightful and earliest commentators were Herman Daly and E. F. Schumacher. Daly's *Steady State Economics* was radical when it was first published in 1973, but it became the underpinning of the field of ecological economics.[10] E. F. Schumacher's *Small Is Beautiful*,[11] published in the same year, has been fundamental to new thinking in developmental economics, especially its emphasis on appropriately scaled technology: big is not always best. Two of my favorite recent books are Tim Jackson's *Prosperity without Growth*[12] and Charles Hall and Kent Klitgaard's highly readable textbook *Energy and the Wealth of Nations*.[13]

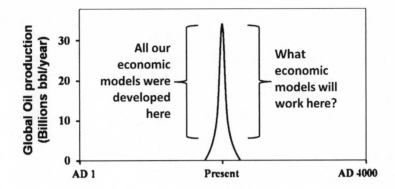

Post-peak economics. All our economic theories work on the assumption of increasing supplies of cheap energy. This "baseline" is not a baseline at all, but a steep upslope. Our economic theories are now obsolete. We need a new economics to face the coming periods of stagnation and decline. *Courtesy of Steve Hallett.*

All these books, and the writings of many others, have explained our predicament with great clarity. Today's world is simply unsustainable, and not just in the sense that it could be greener, but in the truest sense: it cannot be sustained; it cannot last. Most authors give prescriptions for changing our economics in order to make them sustainable enough for us to continue on, but I think the time for that has now passed. Nothing can prevent this system from failing. Living within the limits of energy and resource supply are prerequisites for long-term sustainability, and no amount of tweaking can bring us even close to the reduced levels of energy and resource use we would require.

I wonder when we passed the point of no return on this trajectory to overshoot. I sense that it was always likely that we would reach this point once the Industrial Revolution began. When we started using fossil fuels, we changed the thermodynamics of civilization. When we began to produce oil, overshoot became inevitable. In relatively recent times, the last leap into an inevitable future of overshoot could possibly be marked at the transition from the cautious approach of the Jimmy Carter administration to the "full speed ahead" mode of the Ronald Reagan administration. A fascinating report on energy and resource issues was presented to Jimmy Carter, pointing out the great dangers of a future of environmental limits,[14] and Carter attempted some responses. One symbolic gesture was to put solar panels on the roof of the White House. In speeches, Carter spoke to Americans about the tough times ahead and the need to tighten our belts. This wisdom was received as weakness; Reagan's gung-ho attitude prevailed: and how! One of Reagan's first actions upon entering the White House was to strip the solar panels from its roof. What followed was called a "morning in America," but we were already deep into the twilight.

Perhaps the inevitability of failure is marked even more recently than that with the economic resurgence and massive consumption of the Asian economies, particularly China's. The Asian economic boom has certainly accelerated us down the overshoot path, and it has also placed those rapidly growing countries in the position of being vulnerable to a much more disastrous collapse than before. Theirs is no economic miracle; it's a ticking time bomb.

The lessons of steady-state economics are essential. It is too late for them to save us from a global economic failure, but not all economics is global and the tough times to come need not inflict their worst impacts everywhere. Preparing for the downturn has become essential at all levels of organization, from household to neighborhood and community, and from town and city to state and nation. Each level of organization is strongly influenced by others, and it will be hard for households and neighborhoods to stay afloat if their city fails, but it won't necessarily be impossible. You cannot prevent the collapse of the global economy, but you might be able to protect your own household, community, or city. The time has come to pick your battles.

The lessons of steady-state economics will have their time again, too; indeed, their message is timeless. The coming decades will be difficult and chaotic, but like the fire-prone forest, we will reorganize and regrow. The time will come when the messages of Herman Daly, E. F. Schumacher, and Tim Jackson must be remembered. This experiment in civilization, like all those that came before, will fail, but perhaps it doesn't always have to be this way.

Feedback Loops and Reinforcing Cycles

Free-market economics is hailed as a sort of natural feedback system that sets the best price for goods and ensures their on-time arrival. Government gets in the way, it is said, and markets are best left to follow this natural system of self-regulation. Customers set the levels of demand and the markets will supply. Supply and demand check each other, just like antelopes and lions on the African savanna, and the system will stay in balance. This aspect of free-market economics has a surprisingly ecological feel to it. The problem, of course, is that it operates in isolation. We forget that economic systems sit inside the larger box of the environment. We have spent decades successfully supplying materials to respond to whatever demand exists, and this has fooled us into thinking that there is an unlimited abundance somewhere, mysteriously "out there."

Nonetheless, since economists have adopted the quasi-ecological thinking of supply-demand feedbacks so readily, it ought to be possible for them to understand ecological and system feedbacks in a broader context. That they have not yet done this is probably because the flows of energy and resources through our economies have been prodigious enough to obscure the bigger picture. We have been remorselessly depleting resources from the environment and degrading its ability to provide more, and this has, so far, had only limited or localized economic consequences.

Feedbacks also occur in ecosystems when demand pushes up against supply. Since ecosystems are always pushing up against their limits to growth, their stability is mostly maintained by short-term feedbacks. The decline of lion numbers in the event of a gazelle shortage redresses the balance of nature quickly: the ensuing decline of lions then permits a resurgence of gazelles. Longer-term feedbacks tend to be more dramatic. The fire-prone forest is an example, and the longer it remains in an overly mature, unsteady state, the bigger the fire will be. It is a feedback event that restores resources to the system, but it does so more suddenly.

Natural systems experience feedbacks of very different durations. The control of blood glucose in the human body is a very fast feedback event whose rapidity is obvious to diabetics. At the other end of the scale are geology-level feedbacks, or at least long-duration pulses,[15] such as the gradual leaching of soil phosphorus that will be restored only by dust storms, flooding rivers, glaciers, or the decidedly slow global cruising of tectonic plates[16]—an environmental pulsing that may occur over hundreds of thousands of years (what a shame, then, that we are depleting the deep, rich soils of the world so rapidly). What took tens of thousands of years to build is being wasted in decades.

We sorely lack responsiveness to the depletion signals of the environment even though the signals have been very clear for some time. When the availability of a resource declines, a natural system will be forced to slow down until the resource recovers. Our systems refuse to slow down. They simply seek the resource someplace else, seek efficiency, and seek a substitute. All this does, of course, is spread the resource depletion wider. When the resource-depletion warning sirens blare, rather than listening, we turn them off. We need to start listening, and carefully. Stability is highest when nature's feedbacks are allowed to act quickly and keep a system oscillating in a dynamic equilibrium. But make no mistake: if the shorter feedbacks do not restore the balance frequently and gently, the longer feedbacks will eventually restore it suddenly and dramatically.

One important factor that affects the duration of feedbacks is closeness of its parts and the type of network structure through which they are linked. The feedback between lions and gazelles is rather quick: they are linked directly. Or are they? I must now admit that I have grossly oversimplified this relationship in a way that has probably made wildlife biologists go nuts. In truth, a shortage of gazelles will cause far more effects. The lions might simply chase after antelopes or wildebeest instead. An abundance of gazelles would be a boon not only for the lions, but also the cheetah, leopard, and hyena. The feedback network, even in this apparently simple scenario, is actually rather complex—and still much more complex than I've presented it.

In truth, the lion-gazelle feedback would likely be subject to failure because it is not a short-term feedback at all. Hungry lions prowling around after a diminishing supply of gazelles could easily lead to the gazelles' extirpation. Lions live for quite a few years, and the feedback might not act fast enough to spare the last few gazelles being hunted across the savanna if they were the only thing the lions had to eat. The complexity and diversity of this system is required to keep it stable. Lion versus gazelle will provide very little stability. Lion, leopard, cheetah, and hyena versus antelope, gazelle, wildebeest, and gnu (and this is just the beginning of the complex web), will provide much more.

Diversity and Equality

Diversity and equality are not just the liberal-talk fancies of the politically correct. They are important characteristics of resilient systems. Diversity is an important measure used frequently by ecologists, and it doesn't mean quite what the layperson might think. Diversity is not just a measure of the number of different species in a given area (ecologists use the term *species richness* for that). Rather, diversity is a function of both the number of different species and the degree to which they are represented. A large oak forest with a single beech tree, two elms, an ironbark, an ash, and a smattering of others might have high species richness, but it is not diverse. Neither is a large corporation with two African Americans, a Cuban American, and a Cherokee on staff. Diversity brings strength and resilience to ecosystems because organisms have different characteristics. People with different backgrounds and experiences bring strength and resilience to businesses. Yes, we are all equal, but no, we are not all the same.

It is rather silly, in the biological world, to ask whether one species is better than another because it begs the question, better at what? Is a giant redwood better than a cactus or a ragweed? Sure, in the hills of central California, but not in a desert or a bean field. Is a human being better than a parasitic worm? Not at living in someone's stomach. The living world has evolved its dizzying array of types wherever there was a niche to fill, and intense competition in bounded systems constantly testing their limits has selected for diversity.

The diversity of organisms—the expression of their genetic variability maintained by sex—gives them the opportunity to evolve to fill new niches. The diversity of ecosystems gives them the adaptability to shift and change as conditions change. Species that abandon sex in the pursuit of efficiency are doomed to extinction in the long term. Ecosystems that decline in diversity are liable to suffer much more dramatic periodic corrections, such as disease epidemics, pest outbreaks, and fire. And it's not just the species richness of an ecosystem that makes it resilient. The odd scattering of rare species in an otherwise continuous monoculture does not enable it to bounce back quickly. Resilience comes from having true diversity: a good mixture of well-represented species.

Ecosystems appear to contain a lot of redundancy, and one might ask whether it really matters that a forest is composed of oak, beech, maple, and hickory when they all do basically the same job of collecting sunlight. An efficiency expert would surely choose the one that seemed the most productive, perhaps the oak, and have the others removed. Productivity might increase as a result, but an epidemic of sudden oak death disease would reemphasize the strength that comes with the diversity of nature.[17]

Yet another sign that the modern world is overly mature and losing its resilience is the way it has depleted so many of these more subtle characteristics of stable systems. We lack diversity and equality, we squeeze out redundancy at every opportunity, and we have become disconnected in many ways. Particularly striking is the question of equality. The modern world is dominated by small, wealthy populations in a few rich countries. And it's not just the rather obvious inequality among countries that should concern us. The growing inequality within countries is also alarming. America is a case in point. We are supposed to be the land of opportunity. America, like nowhere else on earth, is the place where anyone can make good. To rise to any position in society and develop abundant personal wealth, all you need is hard work and initiative. America is the epitome of meritocracy. Or do we just keep saying this because it was once true?

Ecologists and demographers use a statistic called the Gini coefficient, which measures the distribution of a given characteristic across a population. Applied to wealth, a low Gini coefficient indicates that wealth is well distributed while a high number indicates that it is not. America has a Gini coefficient of 46.6, one of the highest in the Western world, nearly twice as high as Sweden's or Germany's. In fact, America's Gini coefficient is not dissimilar from Iran's or Mexico's.[18] This should come as no surprise, of course. We know that the richest 1 percent of Americans control half the country's wealth (and most of the top 1 percent are paupers compared to those in the top tenth of a percent). This land of opportunity has come to be dominated by the wealthy and superwealthy, and things have changed rather quickly. America's Gini coefficient was ten points lower only a quarter century ago.

The consequences of the widening gap between rich and poor have been many and varied, and the overall outcome has been decreased social mobility. Those born in poverty have much less opportunity to climb out of it than they think. Those born into wealth have better odds than ever of staying there. The increasing wealth inequality among and within countries is a natural consequence of a system that is approaching limits and reaching increasingly for efficiency, and it is yet another symptom of a system on the brink of failure.

Pick Your Battles

> Hope is the real killer. Hope is harmful. Hope enables us to sit still in the sinking raft instead of doing something about our situation. Forget hope. Honestly and candidly assessing the situation as it truly stands is our only chance. Instead of sitting there and "hoping" our way out of this, perhaps we should recognize that realizing the truth of our situation, even if unpleasant, is positive since it is the first step toward real change.
>
> —Gringo Stars, quoted in Derrick Jenson's *Endgame*[19]

"Pick your battles" is one of the best pieces of advice my mum gave me when I became a parent. My big brother, John, when he was a little kid, was apparently being a bit of a brat one day, refusing to stop playing with one of his toys when he should have been coming to the table for supper. Mum repeatedly asked him and then eventually warned him that if he didn't come she'd throw his toy in the trash bin. Silly John didn't take her seriously and ignored her. Ignored my mum! I know! Hard to imagine. Anyway, the toy went in the bin. John was distraught and Mum was upset, too. She didn't want to throw the toy away, and John was so sad. "You can't control everything, so when you put your foot down be sure to follow through," she said. "Pick your battles."

Our economists and politicians are all headed off on the wrong track. The economic downturn we have experienced over the last few years is yet another sign that we have reached and overshot our limits to growth. Surely the time has come to rein in the engine of growth and try to fix things, but there is no sign of this from our leadership. Our leaders continue to insist that we must revitalize our economies and get them growing again. There are environmental issues, to be sure, and so we promote efficiency as a means to fixing the environment—sort of—without slowing us down. But efficiency is a trap. It doesn't help one bit, and it weakens us in the process.

The list of things that we actually need to do to avoid collapse is impossibly long. All of them need to be fixed, but few of them will be. We need to reduce our consumption of natural resources dramatically, restore our forests and soils, and stop the depletion of freshwater resources, fisheries, and metals. Unfortunately, we also need to reduce our population and increase our production of renewable energy by orders of magnitude, and we would need to do all this on the backdrop of a warming world. What politician is going to come up with policies that would make a serious dent in that? Not one who needs to show progress over the next four years, that's for sure. The time has come to focus on the problems we can solve at a smaller scale. There is not much most of us can do to change the trajectory of society, but there is a lot we can do to prepare ourselves for the future, and we can certainly try to have an influence locally.

What are you going to do to prepare for the future? Buy gold? That's going to be hard to eat if the collapse is total. Run to the hills? What if you get sick? I'm afraid I can't provide specific solutions because everyone's situation is different. What I can say, though, is that the time for trying to save the world is over and the time for making your own world more resilient has come.

Contorting one's mind around all the problems faced by the modern world is something akin to traveling the five-step path of recovery from the death of a loved one, a painful loss, or an addiction. We first deny that there is a problem at all. But our addiction to oil, our disastrous loss of natural resources, and the death of many of our ecosystems can no longer be denied. So we become angry. There are plenty of very real culprits at whom we can direct our anger: corporations, the government, or freeloaders, but we gradually become aware that we are also to blame. It is not just the power company that is responsible for climate change. You, too, bear responsibility for heating your house. So we enter the bargaining phase. We try to bargain our way out of peak oil and climate change by making better purchasing choices and by trying to increase our efficiency. It is in our bargaining, however, that the efficiency trap rears its ugly head. Understanding the efficiency trap can curtail bargaining and send us into depression. There seems to be nothing valuable we can do to solve the great challenges we face. Through this fourth, nastiest stage of recovery, however, we can begin to recalibrate our sights. Finally, we arrive at acceptance. We cannot save the world as it is—nor should we—but with acceptance and understanding comes the ability to take action. We can now find ways to participate.

There are lots of deniers, and many people are angry. The vast majority of people are engaged in bargaining, but in my view, the bargainers are not confronting the real essence of our predicament. The bargainers are the most important audience for *The Efficiency Trap* because they have the will to make changes but lack direction. Many people are depressed, seeing no way to participate, but there are a thousand things to be getting on with. No, we cannot save the world as it is, and we must accept that, but we can soften the blow and prepare for the best possible recovery. Resilience is the key. You are likely to be hit hard by the coming collapse, so you must be resilient enough to transition from this economy to the next. And you will probably not be able to do this on your own. The key to resilience is to strengthen vital systems and build resilient communities.

CHAPTER 12

STRENGTHENING VITAL SYSTEMS

Then you write, "oh, they're just rearranging the deck chairs on the Titanic." First of all, that's a terrible metaphor. This administration is not sinking. This administration is soaring. If anything, they are rearranging the deck chairs on the Hindenburg.

—Stephen Colbert,
White House Correspondents' Association Dinner,
April 29, 2006, Washington, DC[1]

Environmental and Social Triage

I think our situation has become so precarious that it's best to start thinking of it as a triage situation like a hospital emergency room in the aftermath of a natural disaster. The guy with the flu symptoms will have to wait while those with life-threatening injuries are stabilized. When we enter the collapse stage we will lack the resources and services that will be demanded, and it's already too late to fix all our problems in advance, so we need to narrow our focus on the things that are needed most, and first.

There can be only limited prescriptions for preparing because we cannot predict exactly how or when events will unfold. The other problem is that the appropriate responses will be location specific. Are you in a rural area or an urban one, and in which country? Do you have land? Does it rain where you live? Is it cold in winter? Your actions should also be determined by the level of influence you can wield in government. Are you a regular Joe or Josephine, a member of the local council (if not, you might want to consider that), a city mayor, or a US congressperson? (Are you Barack Obama? In which case, "Hello Mr. President, can you rustle up some more Obamacare and a tad more green energy, please?)

Efficiency in Its Place

I must speak briefly on the virtues of efficiency. I know this will come as a surprise, given that I have spent the last hundred pages or so calling it a trap, but although efficiency tends to increase throughput we mustn't forget that it can, at least in theory, deliver the same service for less. The problem with efficiency is that we expect it to save energy and resources while keeping our economies moving. It will not help with these goals, but there are vital systems that we need to keep functioning. We will need to readjust in a range of different ways, eventually finding alternative sources of energy for many of them. The amount of power that alternative energy sources can deliver will be less than we currently enjoy, and so, in conjunction with downsizing and redesigning, high levels of efficiency will be needed in some places. The best example of this is in building construction.

We have chosen the wrong efficiency route over the last half century by choosing efficient construction over robustness and resilience. Most of our houses are cobbled together with two-by-fours (and when did they become one-and-a-half-by-three-and-a-halves?), staples, and chipboard (and did you notice how crappy the timber in the hardware has been getting in the last few years?). We toss a little fiberglass insulation in the walls to reduce heat losses and beef up the heating system with a large furnace. The furnace might be highly energy efficient, of course, in the way it converts natural gas into heat, but who are we fooling? Without natural gas and electricity our houses will become unbearable when the outdoor temperature fluctuates too far from our comfort zone.

Two perennial confusions in discussions of efficiency are the difference between efficiency and effectiveness, and the difference between renewable energy and energy efficiency. We consider a building that gets some of its electricity from a renewable resource to be energy efficient. Perhaps it is, but as likely as not, it is no such thing. Being able to obtain power from more than one source, however, may make it much more effective and resilient. We tend to fall into the efficiency trap with such projects by adding extra comfort features or making the building bigger than necessary, but buildings with lower energy demands and less reliance on fossil fuels make a lot of sense. Whether or not such buildings are more efficient is less important than the fact that they will be more resilient—and resilience should be the primary goal.

Buildings with renewable energy features—generally known as efficient buildings—will tend to be more resilient. A house designed with strong passive solar features, for example, can capture the warming rays of the low, winter sun but block the sun's rays in the summertime, which can reduce the energy consumption considerably. The initial costs will tend to be higher, but well-constructed buildings can provide effective lighting, heating, and cooling with very little energy consumption. They may struggle to pay for themselves in energy savings in today's market, but they will pay great dividends in the future: not because they are more efficient, but because they are more effective. The difference may not be very obvious while we have an abundance of cheap fuel, but it will be stark when we do not.

Efficiency in building construction has an important role to play as long as the focus is on reducing energy needs to a point that might be supportable. Build a small, sturdy, well-insulated house with as many passive features as you can and then, yes, for your supplemental needs, get the most efficient system available.

There is enormous value to be had in efficiency improvements because the problem is not with efficiency itself but with the unintended consequences that flow from it. Efficiency will continue to have unintended consequences, but as we move into a future of declining resource availability and the failure of systems, efficiency improvements will be needed to keep some of our vital systems operating. We need to be careful not to jump back on the bandwagon, but efficiency has an important role to play in developing self-sufficiency, particularly at the household and community levels. Efficient components have a part to play in robust and resilient systems.

Recycling and Down-Cycling

Whereas natural systems are closed-loop recycling systems, our systems are open throughput systems, and this is a sustainability nightmare. We will be forced to reduce our consumption as a number of resources go into decline, and we have thus far clearly failed to reuse and recycle materials sufficiently. The degree of recycling required to sustain a system depends on location, the value of the resource in question, its renewability, and the problems caused both by its production and the recycling itself. Some resources are in finite supply while others are renewable, at least in theory. At the end of the day, however, finite resources must be completely recycled or they will eventually run out, and renewable resources must be consumed no faster than they can renew. I think the most important mistake we make with recycling is to place too much emphasis on the energy that it consumes rather than the materials it recovers.

Pennsylvania Landfill. We are shaving the tops off mountains to dig out coal. Thank goodness we are using our trash to put some mountains back. *Courtesy of Steve Hallett.*

Recycling is one of the glaring differences between ecosystems and economic systems. The recyclers sit at the bottom of the food chain in ecosystems, but they are its most abundant members. A single gram of soil contains millions of microbes from thousands of different species that scavenge nutrients and return them to the rest of the ecosystem. The microbial world lies unseen, but it is the largest and most diverse component of the living world by any measure. As tiny as each bacterium may be, their total mass exceeds that of the rest of the living world combined. Microbes are the last word in the ecosystem's energy cascade, but their energy demands are significant, and a huge proportion of an ecosystem's available energy is consumed at this stage. The energy consumed by microbes is very obvious from the heat generated in a compost pile. Throw together a bunch of dried leaves, grass clippings, and food scraps, give it a mix, make sure it is moist, and then watch what happens as the microbes go to work. The pile will get hot. In the winter, it gives me great joy (weird, I know) to see my compost pile steaming away through little vents it has melted through its covering of snow. That heat is coming from the microbes that are busily respiring as they turn your apparently useless waste into beautiful plant food. Ecosystems invest a large share of their energy into recycling, so they do not run out of resources; we invest very little into recycling, so we will.

Because we can synthesize the coolest things from petroleum, burning it has to be rated as among of the dumbest things we have ever done as a civilization. There is a vast cornucopia of petroleum-based products that could have been used to build a sustainable, vibrant world, and we had huge petroleum reserves not that long ago. We chose, instead, to set most of it on fire. Most of what was left was used to synthesize crap we didn't need, disposable crap that was toxic to the environment, was used once, and then was dumped in a hole in the ground. Instead of using petroleum to build a sustainable future we burned most of it as fuel and used the rest to build huge plastic mats and trillions of little plastic pellets that will float around in our oceans for centuries as a reminder of our missed opportunity.

The opportunity for using our petroleum endowment to build a sustainable future has passed, however, and it doesn't help to pretend that it hasn't. We have used roughly half our endowment of fossil fuels, and we've developed an addiction to burning them. (We shouldn't be surprised: we were addicted to burning whales before petroleum came along. . . . We seem to like setting stuff on fire.) The vast majority of what remains will be burned, and the rest will be used to make more crap, so I don't think we should be too squeamish about burning some of it to salvage needed resources. Critics claim that recycling is energy intensive, and in some cases it is, but it is generally less energy intensive than making new products.[2] Recycling is an excellent use for our energy resources.

Recycling will help to keep plastics out of the environment, it can reduce the amount of mining needed for metals, and it can reduce deforestation and water pollution caused by the paper industry.

Recycling is on the increase again in most of the developed world, and this is encouraging because recycling will become much more important in the future, especially for materials that are energy intensive to produce from scratch or are reaching limits, such as many metals. Our recycling efforts remain insufficient, however, and we tend to have effective recycling programs only when there is a clear economic advantage. Huge volumes of useful materials still go into landfills. We can look at recycling programs from the Second World War for some encouragement that it is possible to ramp up recycling efforts when the need is high. Strong recycling programs were driven by government mandate in many countries during the war. It is discouraging that they were largely dismantled when the flows of materials were later restored.

"Aha!" I hear you say, "but what about the rebound effect?" Yes, well spotted, there is a rebound effect. If recycling reduces price, it will increase throughput. In this way, effective recycling can be a factor in economic growth. And, yes, there is also an indirect rebound. Cheaper plastic will enable the production of, say, cheaper computers that also contain silicon, copper, and many other materials. The advantages of recycling override these rebound effects, however, and have their greatest impact in curtailing system effects. Most importantly, recycling is a resiliency strategy, and it is something we will need to be very good at in a few years. A bit of practice, and having effective systems in place, can't hurt.

James Howard Kunstler, the author of *The Long Emergency*, one of the earliest and best books on peak oil,[3] has written a couple of postapocalyptic (post-peak-oil) novels. In the first one, *World Made by Hand*, he picks up the lives of a bunch of people trying to survive in a destitute post-peak-oil town in upstate New York. They have adopted various classic strategies to figure out a path into the future. There is a land baron with a community of serfs, a rather creepy missionary group, and the regular townsfolk, trading veggies, cheese, wood, and other basic necessities. And then there is the biker gang. The bikers have risen to prominence by taking over the municipal dump. If you want any hardware, say a box of nails, you have to buy it from them. In Kunstler's post-peak scenario, the richest pickings are in the landfill. It's highly entertaining stuff, and it has a disturbingly realistic feel to it.[4] Let's hope recycling does not become as necessary as this.

Better by Design

> If the first Industrial Revolution had a motto . . . it would
> be "If brute force doesn't work, you're not using enough of
> it." . . . The standard operation instruction seems to be "If too
> hot or too cold, just add more fossil fuels."
> —William McDonough and Michael Braungart,
> *Cradle to Cradle: Remaking the Way We Make Things*[5]

Recycling is essential for the recovery of materials back into the system so that the system can be as closed a loop as possible. One of the ways we have drifted into such a high-throughput ethic is with the products we make. It has become increasingly efficient to produce goods that are neither durable nor recyclable. There has even been a trend toward designing goods specifically so that they will need to be replaced early in order to increase sales, and this is known as "built-in obsolescence."

Poor old Mum had to buy a new washing machine a few years ago. She'd had the old one since she was married in 1958. Mr. Quarmby came by every so often to do routine maintenance and keep it running, changing the odd hose here and there, and run it did, for forty years. But the machine eventually started to show signs of wear and tear, having washed nappies and grass-stained knees for three scruffy kids nearly every day. Eventually, Mr. Quarmby told Mum it was time to let go of it. She deserved a better, more efficient machine. Mum finally caved and got a fancy new washing machine. "It looked prettier than the old one and had all sorts of fancy wash cycles," she says, "but it didn't wash the clothes as well as the old one." It also had a fancy tumble dryer so that she wouldn't need to hang the clothes out on the line to dry anymore. (Although she did: I'm not sure she ever used the tumble dryer.) The fancy new machine, of course, was kaput in a few years and Mum had to get yet another. "Wish I'd kept the old one," she says, "a bit of soldering and it would still be working fine."

A similar thing happened with the old cooker. The fancy new one lasted a few years before the glass door shattered one day. Not worth fixing. New stove needed. She does still have her old fridge, which she inherited from Grandma in 1977. "Now't wrong with it," she says. "I expect it'll see me out." It's a tiny little thing about the size of a hotel minibar. I couldn't even fit a week-end's supply of beer in the thing, but Mum has been quite content with it for thirty-five years. My fingers are crossed that the fridge will keep on going. Incidentally, it "wastes" more electricity than mine, but it consumes far less.

We need to get back to the ethic of making things to last. Robustness is not the same as resilience. Robustness is the ability to withstand shocks without damage whereas resilience is the ability to bounce back despite shocks. However, a system with more robust components is likely to be more resilient. We also need to get back to the ethic of designing things such that they can be fixed. Those fancy computerized gadgets in your car are nice comfort features, but they make it much harder to repair. Far fewer people tinker with their cars these days. I remember helping my dad check the points and plugs on his old Ford Cortinas (he had three of them in a row over a span of about twenty years). Cars don't even have plugs anymore. When things are not right with your car today its computer will flash that annoying "check engine" light and off you go to the garage to find out, for the seventy-dollar diagnostic fee, that the sensor is broken. There's nothing wrong with the engine at all, and the replacement sensor will cost you $200. Joy. And what happened to all the decent mechanics?

We also need to improve the salvageability of our products, or at least their recyclability. An excellent book on all this (which is printed on super-durable, synthetic "paper") is *Cradle to Cradle* by William McDonough and

Michael Braungart.[6] *Cradle to Cradle* explains the many ways in which products can be designed to last, salvaged, reused, repurposed, recycled, or downcycled. And it encourages us to be much more cognizant of the full life cycle of products. *Cradle to Cradle* is a response to our existing cradle-to-grave ethos, and it asks how we can design products better to help reduce the outflow of our precious resources from our economic system. McDonough and Braungart point out that we need to close the loop in order to establish a sustainable economy.

We have needed to be much better at all this for many decades, and our failure to stem the outflow of vital resources has brought us to the brink. Improving these practices can no longer save us from failure, but it can play a vital role in softening the landing and seeding the recovery. Working on these things now will give us some practice, at least, and doing so will put the systems in place for when we need them, not just to improve our green image, but for our very survival.

Protecting Wild Lands

> If a man walk in the woods for love of them half of each day, he
> is in danger of being regarded as a loafer; but if he spends his
> whole days as a speculator, sheering off those woods and makes
> earth bald before her time, he is esteemed an industrious and
> enterprising citizen.
>
>   —Henry David Thoreau, "Life without Principle"[7]

Each burst of civilization and technological advance provides us with a moment in the sun but lessens our planet. The first Americans, armed with stone-tipped weapons, drove the great American megafauna to extinction. The glyptodont, saber-toothed tiger, mastodon, ground sloth, and cave bear will never be seen again. The Mesopotamians turned the Fertile Crescent into, well, just the crescent. The ancient Greeks and Romans desecrated the once-fertile Mediterranean basin, much of which remains rocky and degraded to this day. The forested wilds of Europe were pushed back into the hills and the far north, as were the North American forests, which the pioneers leveled like a colony of army ants moving across the landscape. By the time this latest burst of civilization has played out we will have sent thousands of species into extinction, destroyed biomes, and created a legacy of degradation that will take nature millennia to repair.

Simply put, we must save what we can. The pressure on our wild lands is mounting, and they will experience a last burst of intensity as we scratch and claw to survive when the rot sets in. Firewood will be at a premium when it becomes one of the best options for heating fuel, as just one example, and we can see what desperate firewood scavenging can do at the advancing southern front of the Sahara Desert. The calls for oil drilling on the US continental shelf, in the Arctic National Wildlife Refuge, and elsewhere will become increasingly shrill.

Two countries that stand out for having destroyed huge areas of unimaginable beauty are China and India. What has happened in these countries over the last half century is an unparalleled disaster. They have pushed and pushed to produce food for their exploding populations and to sustain economic growth that has raised them out of poverty. The remarkable success they appear to have had has merely widened and deepened the trap. Economic growth has caused ecological carnage that will become economic collapse. China and India will descend into poverty and chaos, millions will die, and the only enduring result will be a depleted environment. We must do whatever we can to prevent our environment from being put to "good use."

There does not need to be any special reason for this. Saving wild lands for future generations is reason enough, but it is also a matter of our own welfare. The array of ecological services that intact ecosystems provide to society is beyond our capacity to measure. Consider the ecological water purification service that nature performs for New York City. The sun lifts water from the ocean into the air: a free pump. Clouds drop the water over the Catskills to percolate down the mountain slopes: a free filter. The city collects clean water. We grossly undervalue the many services such as these that natural systems provide. Putting a monetary value on all this is an exercise in futility, but imagine New York City having to get its water from desalination plants and you will gain some vague understanding of the value of this one ecological service.

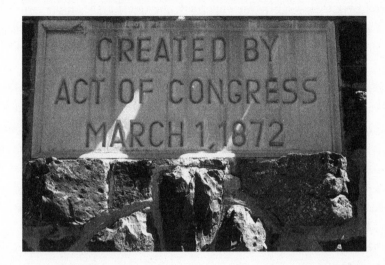

Yellowstone was America's first national park, created by an act of Congress during the Theodore Roosevelt administration. *Courtesy of Steve Hallett.*

Power to the People

> We should never doubt that a small group of thoughtful, committed citizens can change the world. Indeed, it may be the only thing that ever has.
> —Attributed to American anthropologist Margaret Mead (1901–1978)[8]

We are going to face a serious energy crunch in the coming years because we will be unable to fill the void from declining fossil-fuel resources. There will be society-wide shortages of electricity coming in the form of increasingly frequent blackouts. The cost of liquid fuels will increase, and then they will be rationed.

We are used to having our electricity supplied over a broad grid, and this has been an efficient way to distribute electricity from large, centralized plants. It does not, however, give individuals or local communities much control over their supply. An important question for the future, when the rolling blackouts begin and the lines at the gas stations grow, will be who will get the electricity and gasoline. An even more important question will be whether or not it will be you.

The transformation of our civilization through the Industrial Revolution and the petroleum interval was driven by access to fossil fuels that delivered energy in a concentrated form, and one of the key societal transformations was that our energy distribution became concentrated, too. The centralization of energy has been one of the most systemic changes to occur as our societies have modernized. Household and local energy generation has been gradually replaced with larger and increasingly efficient centralized distribution systems.

Your lights access electricity generated hundreds of miles away at the flick of a switch. Your heating system accesses natural gas that may have traveled through a thousand miles of pipes from field to furnace.

It is not necessarily the source of the energy that we recognize as modern; more importantly, it's the low cost and convenience of the energy delivery. We want access to energy without having to deal directly with its production. It is for this reason that we want to keep the system intact. We recognize that there are problems with fossil fuels as sources of energy, and we would like to replace them with renewables, but we seldom consider that renewables might not work within the existing centralized system.

Renewable energy technologies such as wind and solar will be difficult to scale up and feed into the grid from remote sites, and the logical solution is to keep them localized. Decentralized energy systems may be very valuable for keeping services running in the case of grid failures.

There are lots of opportunities for heat and electricity cogeneration at smaller scales. We tend to get our heat and electricity from separate sources. However, systems that collect or generate heat and electricity locally will be much more efficient and effective in many cases. Strategies like this will be increasingly important for rural towns whose supply will be in jeopardy. The time to start erecting some of these systems is now, even though their return on investment is not yet obvious.

A number of communities have developed innovative community energy projects using a range of ownership models. They have often needed to rely on government grants and other forms of support, but they've begun to demonstrate that much can be done by innovative communities.

The borough of Woking in the United Kingdom has been a striking success, and Woking is a reasonably large town of about seventy-five thousand people.[9] All its municipal energy needs, and much of its residential needs, are now delivered completely independently of the United Kingdom's national grid. Woking installed a series of distributed heat and power cogeneration plants that convert natural gas to electricity and heat with hydrogen fuel cells. There are also photovoltaic (PV) cells on many roofs in the town, and they've also installed innovative features such as solar PV-powered parking meters. Solar energy generates electricity during the day and the cogeneration plants provide it at night. Decentralizing has reduced electricity costs in Woking, and the town's resilience has increased as a result.

There was resistance to the development of the project from the entrenched power companies, and Woking was not permitted to use existing power lines. They installed their own. The Woking model was an early demonstration of the fact that decentralized power can work, and it has continued to work effectively since the early 1990s. The biggest weakness of the system is its reliance on natural gas, which is something Woking will need to get away from at some point, but I'm betting they can do it.

There are now thousands of communities around the world that have taken themselves off the grid. The most common mechanism has been through the use of community wind projects.

The small towns of Daylesford and Hepburn Springs in Victoria, Australia, built the Hepburn Wind Project, consisting of two turbines, on a nearby hill.[10] The island community of Samso, Denmark, population four thousand, went off the grid by adding ten offshore turbines, and the development of decentralized solar and biomass facilities has made its residents nearly independent in terms of heating and cooling.[11]

Projects such as these are becoming increasingly common as communities find their own particular models for utilizing community energy, whether it is with the express goal of building resilience or to export electricity to the grid as a business strategy. Most of these projects remain small, and success has been achieved mostly within rather tightly knit communities (further emphasizing the need for community cohesiveness and cooperation). Much less progress has been seen in larger towns, and an example of the size of the needs in larger communities is evidenced by City Bus in my home town of Lafayette, Indiana.

Kudos to City Bus, which has erected three wind turbines that offset about 180,000 kW/h per year of their energy needs. It's a great project, but it also gives a measure of the transportation problems we will face in a world without oil. City Bus is a transportation company (right: you guessed that), but its turbines can cover only the electrical needs of its buildings. City Bus estimates its energy savings as "14,000 gallons of diesel equivalent," but that's a bit disingenuous because the wind energy replaces electricity, not diesel. To replace the diesel it would need to switch to EV or hydrogen buses (with the associated technical issues) and it would need roughly another seventy turbines to do so.

City Bus demonstrates that our energy future will require much more than simply supplementing fossil fuels with other sources of energy. Many of our services will need to be redesigned and delivered differently. Transportation, in particular, will present huge challenges. After all, if seventy wind turbines are built in Lafayette, it's unlikely that they will be for powering the bus company.

Renewable energy technologies can be particularly effective at the scale of the individual home. This is not always the most efficient way to deploy them, in terms of strict cost and energy conversion factors, but it might be the most effective. It can give the homeowners control over their supply, and they can match it to their needs. Emergency lighting can be supplied by a very small, simple, and cheap system. Running a few lights and a refrigerator (perhaps your immediate needs) is not too much of a problem. Powering an entire house will take a decent-sized solar array, the scale of which will depend on your demands. Running the air conditioner nonstop in a large suburban McMansion may not be an option. Hot water can be delivered very cheaply and effectively by quite simple solar water heaters. Home heating and cooling can be delivered by slightly more elaborate underground geothermal systems if you have a little garden space: you need very little electricity to power that, and heating can be supplied from various sources. All in all, it is quite possible to prepare many homes, especially small homes, for energy shortages with a well-thought-out combination of insulation, low-consumption appliances, and a modest amount of home-grown renewable energy, and this is a project to be starting on now while there are subsidies for many of these things.

There are a number of problems with renewable energy, but there is significant potential if it is deployed the right way. The approach of seeing wind, solar, hydro, and biofuels as replacements for fossil fuels within the current system is problematic. The hole that will open up in our centralized energy system cannot be plugged with renewable energy. It will become necessary for local governments and individuals to fill their own energy gaps.

Since one of the biggest problems with renewable energy is that it comes in a dispersed form, it makes sense to deploy it in a dispersed way. Large wind and solar farms are not a great investment. They have a rather poor energy return for both the energy and capital invested, which is why investment has been modest. Much more effective than trying to develop large, centralized renewable energy projects would be encouraging lots of investments in smaller projects. Subsidies and tax breaks are available, and I would encourage communities and households to take advantage of those while they are. I would also advocate for increasing government support to local energy projects. One initiative that has been particularly successful is the subsidization of solar photovoltaic cells and the modification of rules so that owners can feed electricity back into the grid. Such subsidies make small-scale solar installations cost effective in the short term, which has increased their rate of proliferation. These systems may not make homeowners free of the grid, but their cells are now installed and can be matched with battery systems if necessary.

The way to encourage the adoption of more renewable energy across a nation is to mobilize the nation's people into decentralizing its energy supply to suit the energy sources that will soon become necessary. Don't get me wrong: it won't be efficient, and it won't prevent the overall collapse, but it will soften the landing for those who have alternatives in place early. Households and communities that develop local energy systems will be much more resilient to the energy shortages that are coming, as will nations that encourage local energy solutions widely.

We apply our best minds and most advanced technologies in an attempt to make wind energy power our modern systems, but the most effective and reliable systems are still simple ones like this. It slowly and methodically pumps water. It may be inefficient, but it is very effective. *Courtesy of Steve Hallett.*

Solar water heaters are very effective—and efficient—because no electricity is involved. The system is simple, cheap, and has no moving parts. Water is heated, stored, and used downstairs. Solar water heaters such as these in Xi'an are very common in China. *Courtesy of Steve Hallett.*

The End of Suburbia: Ghetto or Transition Town?

Challenges will be faced everywhere and by everyone in the coming decades, but some places are inherently more resilient than others. A well-connected rural community in a diverse and productive landscape might be particularly resilient. Many of our rural communities have decayed badly over the last half century because there is not much potential in these communities for high-flying, well-paid jobs. Others have been converted into bedroom communities, which is a real shame. While rural towns tend not to flourish in the boom times, neither do they collapse as readily during a bust. At the other end of the scale are some obvious postapocalyptic nightmare scenarios. Cheap energy has been the lifeblood of modern-era boomtowns like Houston, Texas, and Las Vegas, Nevada. Houston is an overgrown, universally air-conditioned oil town with a huge wealth gap and a serious lack of local resources (other than oil). Las Vegas is in the middle of a desert (in case you hadn't noticed), and it depends on armies of people rolling in on cheap jet fuel to spend the spare cash that they will soon not have. If you live in one of these places and are looking to make your situation more resilient, your first goal is clear: get out.

A number of towns around the world have studied future trends, recognized the threats, and made specific and coordinated efforts to transition. I encourage readers to look into the Transition Network,[12] and there are some great books available, too.[13] Transition initiatives and transition towns attempt to go further than community power projects by examining needs for all services, not just energy. They emphasize the building of resilience against the twin challenges of peak oil and global climate change, and their focus tends to be more upon food production rather than on energy. Other innovative concepts such as local currencies have also been adopted. Totnes, a town in the United Kingdom, has the "Totnes Pound," which can be traded at many local businesses. The inspiration for transition towns came from the broad concepts of permaculture that were first developed by Australians Bill Mollison and David Holmgren.[14] The transition movement was initiated by permaculturalist Rob Hopkins, and the first transition towns were Kinsale, Ireland, where he taught at the local college, and then Totnes, UK, to which he later moved.

The idea is relatively new. Though most transition towns are in the relatively early stages, some of the planning documents they have prepared, often called Energy Descent Action Plans, provide extremely valuable insights to their communities and city leaders, showing the way forward into a new economy.[15] One strength of the transition movement is its thoughtful involvement and mobilization of the community in planning, so transition initiatives tend to be shaped by local needs rather than just the copying of models that may have worked somewhere else.

Probably the most interesting target for resiliency analysis is the suburbs. Millions of people live in these rather odd "communities" that are neither country nor city. In some ways they reap the benefits of both country life and city life. On the positive side, suburbanites tend to have at least a smidgen of space to work with, and they may have some spare cash that they can use to make preparations for the coming collapse. There is a lot on the negative side, however, and the suburbs also represent the worst of both worlds.

Suburbs can only rarely be considered communities, and they are most definitely not towns. They do have some characteristics of communities and towns, however, and the big question for the future will be which suburbs can be developed into towns and which will become ghettos? A lot, I think, will depend on how suburbs develop over the last few years of cheap energy that we have left.

The suburbs are a construct of the postwar, oil-powered economic boom. They were created by the desire of people to get out of the dirty, industrial city cores to find a quieter, more pleasant environment in which to live. Land was cheap, and traveling in and out of the city was streamlined by the automobile. People moved in droves and the countryside was replaced with countryside "mock-ups" characterized by front-facing fake window shutters overlooking a manicured patch of lawn.

Suburbia took its peculiar form because of when it happened, and because it happened so rapidly. It's as if the petroleum generation was scurrying out to the burbs for the express purpose of giving birth to the massive baby-boomer generation. The large concentrations of people moving to new developments outside the city created natural locations for new businesses, but the businesses that arrived were of a particular type. The new suburbanites were a population of drivers, and so commercial developments were designed to accommodate the automobile.

It began with the shopping mall, which brought the butchers, bakers, and candlestick makers all together in one place. The shopping mall is actually a fairly old concept, and some of the earliest ones were upscale centers in the shopping districts of older cities or extensions of covered public markets. The Burlington Arcade has been selling very swanky stuff in Central London for two centuries. Tommyfield Market in Oldham (my corner of Manchester) has a centuries-long history, became a covered market in the 1960s, remains in the center of town (rebuilt after a fire in the 1970s), and is now a combination of market and shopping center. This long history of shopping centers has helped European cities retain a diverse retail core. The biggest shopping center in Manchester, the Arndale Centre, is right in the middle of the city.

The mall took a very different turn in America during the exodus to the suburbs. Newly constructed malls put the stores together in a big, efficient block, and big-box stores arrived and planted themselves next to the hordes of newly arrived, gainfully employed home builders. There was not the slightest need to make either the malls or the big-box stores accessible on foot or by public transport, so they were plonked down with large parking lots on cheap land. The same thing happened with many other services. Out went the small, quaint city-center movie theaters (there may be one or two left, playing indie films) and in came the drive-ins and then Cineplexes. That there are not many sidewalk cafes in the suburbs should be no surprise given that there are not many sidewalks. No problem: eat at the food court or stop at the drive-through.

So how do we go about preparing the suburbs for a future when big-box stores start to move out, commutes becomes impossible, and jobs in the city dry up? How do we give these sterile, auto-centered suburbs a chance to develop into thriving towns rather than dilapidated poverty traps? Well, if you want to do an analysis of your suburb, try abstaining from cars for a couple of weeks and see how it goes. Travel on foot, by bike, and on public transport, and see what you struggle with. At the same time, look for opportunities: Where should the new bike shop go? What about the vegetable stand?

For the most part, building the resilience of the suburbs will be a process of reversing the trends that were put in place for efficiency in an era of cheap energy. The goods offered by Walmart and the other big-box stores will need to be offered once again by mom-and-pop stores. The jobs in the city will need to be brought closer.

The suburbs will not mysteriously morph into thriving towns just because the need arises. Some might do rather well with relatively little preparation, but others are in desperate need of a makeover. Some of the things that can be done to make suburbs more livable and capable of transitioning will be things like adding bike paths, urban gardens, and farmers' markets. Heck, even a few sidewalks would be nice. Many suburbs are so car dependent that they don't even have a sidewalk where you can walk the dog, much less walk to a store. Let's put some sidewalks in to give a little incentive for the stores to come. Plant some trees. How about a park where people can meet? A community is not a community unless a few people in it actually know each other.

And what is it with that ridiculous suburban front lawn? It's great to have a nice front yard with grass and some flowers, to be sure, but we seem to have become a little obsessive about it, so much so that councils often have complicated laws about what you can and can't grow in front of your house. The grass must be mowed to some particular length, and so on. It is quite possible that you are not allowed to grow food in your front yard.

The city of Ferguson, Missouri, cited resident Karl Tricamo for his "unsightly" front yard and demanded he tear it up and plant a lawn. This type of nonsense happens all over the country. Tricamo, it turns out, is a pretty good urban farmer. He grows dozens of different vegetables and herbs in his front yard, providing the majority of the vegetables his family eats for a good portion of the year. He grows summer crops and winter crops, and he mulches over the offseason. He also does a lot to make the garden attractive, planting in neat rows and arranging the plants to be appealing.

Tricamo went to the city's Board of Adjustment and won, but here's the kicker, and it says a lot about our attitude to food and the environment. Even as the board delivered a 4–1 vote in favor of Tricamo, one of the councilor's made the following comment: "The board felt that, technically, he had the law in his favor, but I think that all of us on the board agreed that the garden is an eyesore. It goes against common sense, really, to put a garden in the front yard instead of the back."[16] Right, and that ridiculous little strip of grassy pretend countryside is what represents common sense? These attitudes are going to have to change—which reminds me of an Eddie Izzard skit on the arrival of the Pilgrims in America (where they were disappointed to find the land inhabited by a native population).[17] "'No, we don't want any of your food, thank you very much . . . just put some clothes on!' Meanwhile, that winter: 'Excuse me. Do you have any food?'"

And if you're looking for new pets I'd recommend raising chickens in the backyard. Again, there's a good chance that's against the law. Get that law changed.

Many suburbs are going to be very difficult places to live as the economy dwindles and fails. The suburbs enjoy some of the advantages of both city and country today, but they are neither one. Preparations are needed to make them as livable as possible in a postpetroleum world.

Building Sustainable Food Systems

> The nation that destroys its soil destroys itself.
> —Franklin D. Roosevelt, "Letter to All State
> Governors on a Uniform Soil Conservation Law"[18]

America produces vast amounts of food, so much food, in fact, that a good proportion of it is sold overseas, particularly soybeans to China, or converted to biofuel. More than a quarter of our prodigious corn crop goes into gas tanks. We have vast acreages of land, much of it high quality (despite quality declines), and our agriculture is technologically very advanced. Yields per acre have never been higher. Despite all this, I am very concerned that America may suffer food shortages in the near future. These shortages will pale in comparison to what most of the world will face, and in America they can be avoided or fixed (although they probably won't be). The world faces a food crisis because food production will decline everywhere and many still-growing populations are in overshoot. Millions are going to die of starvation and malnutrition, especially in Asia. Food production will decline and food prices will rise when diesel (oil) and fertilizer (natural gas) become prohibitively expensive. The disastrous depletion of soil and water resources will become abundantly clear when we cannot afford to prop up production artificially.

The United States is in better shape than most countries because its population is not in serious overshoot and the quality of its land remains relatively high. The rising costs of fossil fuels will make it impossible to work the land in the way we do at the moment, but this a good thing in many ways. Picture American agriculture in a future without fossil fuels: no big machinery to pull plows, planters, and harvesters, no nitrogen fertilizer (and no phosphorous either due to rock phosphate depletion), and fewer pesticides. The transition needed in order to reach that future is huge, and the bottom line is simple. You can't farm a thousand acres without those big green-and-yellow toys. Farms will need to shrink and millions of people will need to go back to the land to farm. It's not going to be easy, but there will be necessary work for millions of people and there is a good chance that the transition can be made. The struggles of American agriculture will cast a big, dark cloud, but the cloud will have a silver lining that could lead to a much more sustainable future. It would be good if we could begin early.

As it turns out, the transition has already begun. What began in earnest with the organic farming movement has blossomed into a diverse array of alternative and sustainable agriculture movements, many of which are proving very successful. The organic movement itself has lost its way somewhat, and much of it has devolved into another form of unsustainable, big agriculture, but it does retain many resilient traits, and its flourishing has demonstrated that much is possible.

There is no single, sustainable agriculture "movement" as such. Rather, there is a great diversity of mavericks of one kind or another with overlapping goals. I spoke with a group of small farmers at a meeting at Purdue University and was struck by how little they had in common. What they did have in common, though, was also striking. They all thought the current model of agriculture was broken. They all wanted to farm the land in a way that would leave it, especially the quality of its soils, in a better condition than they found it. They all wanted to produce healthy food and make a profit. And they all wanted to share their vision and experiences. Many people have invited me to visit their farms, and they don't close any doors. These farmers will walk you through every greenhouse, chicken coop, and shed. They will show you where they get their water and how they mix and spray their (usually nonsynthetic) pesticides. They will tell you how to cover the broccoli with netting to keep the pests off, when the pests arrived every year for the last five years, and why brussels sprouts should be left in the field until after the first frost (so that they taste better).

Clay Bottom Farm is a modest but successful business near Goshen, Indiana, owned by Ben Hartman and Rachel Hershberger.[19] It is a five-acre property with just two acres cultivated. Yes, that's less than 1 percent of the acreage a corn-soybean farmer is supposed to need (and Ben and Rachel actually grow food). They use no fertilizer at all but instead compost the land heavily. The organic content of their soil has, in just a few years, risen from less than 2 percent back to the whopping 10 percent that it probably had before the first farmers arrived centuries ago. The soil holds water and nutrients better every year and less compost is needed. And the crops look fantastic.

The compost, by the way, is made from municipal waste that would otherwise be discarded.

Rachel and Ben sell through three routes: at the local farmers' market, to local restaurants, and with CSA baskets that people pick up at their farm.[20] It's all very local, very sustainable, and very simple. What is not simple about Clay Bottom Farm, however, is the team of Ben and Rachel. They're really smart, and not only smart farmers who locate and rotate their crops in careful patterns to preserve fertility, suppress pests and diseases, and monitor for pests and diseases like tax auditors; they are also smart businesspeople. They have done a sort of Six Sigma analysis of their business to reduce waste (a zero-muda farm!), ensure high quality, and maximize profit. They grow what works and skip what doesn't. Sweet corn is a pain in the neck, collects ear worms like an unbathed kid collects lice, takes up a fair amount of land, is greedy for nutrients, and sells for virtually nothing. Simple solution: don't grow sweet corn. Baby greens, on the other hand, take up very little space, are ready for harvest very quickly, and sell for a good profit at restaurants.

Clay Bottom Farm has a baby-green production system that would put the Toyota Production System to shame. A custom-modified planter pops the tiny seeds into a meticulously prepared seedbed like perfect phalanxes of miniature soldiers. The greens are harvested within as little as three weeks, washed, and then spun dry (they store much longer with the moisture removed) in—get this—a basket mounted atop a reclaimed washing machine that runs on its spin cycle. The greens are then sealed in a bag and put as soon as possible into a homemade cold room. The tiniest section of the farm, just a handful of forty- and sixty-foot-long rows, delivers a steady income stream. Clay Bottom farm will, I am sure, do extremely well through the coming years of transition, and it is making its local community more resilient as well.

The local vegetable farming guru in my neck of the woods is Kevin Cooley of Cooley Family Farm, which is located just outside Lafayette, Indiana. His story is similar to Ben and Rachel's. Kevin grows an abundance of food on just ten acres. And this guy gives back to his community. He will speak to anyone or any group that wants to know how to produce healthy food, build soils, and protect and sustain a piece of land.

Neil Moseley is another type of innovator. He's part of a large farming family south of Lafayette that has grown corn and soybeans and raised hogs for many years. Neil, however, is definitely a fully paid-up member of the Maverick Farmer Society; he wanted to do things differently.[21] He took over a couple of old hog barns, cleaned them out, removed all the stalls, ripped off the roof, and replaced it with clear plastic to make greenhouses. (Some academics, by the way, told him it wouldn't work: it worked just fine.) He then cut a deal to collect corn stubble from the rest of the family's acres and invested in a corn crusher and biomass burner to heat the greenhouses. He can produce food year round. Want a head of locally grown lettuce in Indiana in February? Neil has one for you.

Ben Hartman and his wife Rachel Hershberger make a living farming five acres in northern Indiana, only two if which are cultivated, defying the conventional wisdom that a thousand acres are needed to support a farming business in the state. *Courtesy of David Johnson, Johnson Creative Photography (http://www.johnson creativephotography.com).*

One last local maverick who deserves special mention is the inimitable Adam Moody of Moody Meats and Lone Pine Farm, just northwest of Indianapolis. Adam is the epitome of "don't tell me it can't be done." I say "inimitable," but I sincerely hope that's not true, because Adam has done something remarkable that I hope will be imitated. Those millions of animals that were once raised on family farms all over Indiana but were gradually carted off to feedlots out West because it was the efficient thing to do: Adam has brought a bunch of them back.

They say that you need a thousand acres to make a profit from farming around here, and what's more, you need much more land to raise animals than plants. So how is it that Adam and his family do very well raising some of the best meat products in the country on 250 acres right here in the great state of Indiana? Adam is a virtuoso in explaining the efficiency trap in agriculture, and it goes like this: Farmers have been seduced by efficiency and the constant push toward increasing yields as a way to increase their income. But while yields have gradually increased, so have the costs of inputs. Meanwhile, food prices have gradually fallen. Farmers have continued to respond with more inputs to further raise yields, and they've also invested in bigger machinery and more land. Their margins have stayed just as slim as ever. It's a disastrous business strategy. Efficiency has promised more and more from less and less, but in financial terms, farmers have suffered the exact opposite. They have found themselves in the embarrassing position of searching for more capital, more land, and more efficiency while the profits disappear down a hole faster than they can dig. Perhaps they should quit digging. What farmers should aim for is the exact opposite: higher margins using fewer inputs on less land, and a system that is sustainable.

Adam's business strives for sustainability at all levels. He is careful to protect the environment and reduce inputs. He raises cattle, pigs, sheep, chickens, and all their feed. (He's getting a dairy cow for when Indiana finally softens its laws against the sale of raw milk—another example of rules that were put in place to shore up the oversized, industrial producers that caused the problem and shut out the small producers that don't have the problem).[22] The key is rotations: everything moves. Chickens are good at chomping up bugs as they wander over a pasture, and they fertilize as they go. Cattle will destroy a pasture if they stay on it too long, but if they are there for a short time they mow and fertilize, as do sheep (a slightly different model of lawn mower). Pig snouts, on the other hand, behave like plows and can cultivate a piece of land for you.

Meanwhile, on the crop side, Adam grows a modest amount of corn, soybeans, and various small grains as extra feed for the pastured animals. An indication of the power of his rotation is how little fertilizer his corn needs. At about a quarter rate once every seven years, his fertilizer inputs for corn are roughly a thirtieth of those of his neighbors. His feed costs are nearly nil (a little extra high-protein feed for brooding chickens and a few supplements): orders of magnitude lower than a feedlot.

Adam's economic model reverses the trend in which most farmers have been trapped. Every input, every hour worked, and every acre owned pays back the attention he pays to it. He value-adds. He runs his own slaughter-house and stores and dry ages his own meat. He smokes his bacon, makes amazing Andouille sausages, and, to top things off, he retails all his own meat out of the Moody Meats shops in the wealthier parts of Indianapolis. The meat is expensive, to be sure, but it's worth every penny—and every penny can be traced back to careful stewardship of a modest piece of land.

A farming business like this is also great for the local economy. He does not cut out the people from the land with efficiency. Instead, he supports about thirty employees. And last, but by no means least, here is a true family farm, passed from generation to generation, prepared and improved by each for the next.

America does not have to rely on vast volumes of oil and natural gas to make food. It does not have to live on fast, cheap calories that cause obesity and diabetes. It does not have to get more efficient to survive. It has to get smarter. Dedicated, hard-working, ethical, and ingenious farmers like these show us it can be done. The transition in agriculture has begun and the knowledge we will need is becoming available.

Another very promising trend in many countries is the reemergence of urban agriculture, community gardens, and backyard gardens. There are some great and fascinating organizations involved in this movement, such as former pro basketball player Will Allen's Milwaukee-based Growing Power, and the P-Patch gardens that began in Seattle, Washington, and have now sprouted in many city spaces.[23] Urban agriculture has a long history; it tends to surge when food prices rise, with notable examples being the victory gardens in the United States and the allotment movements in the United Kingdom during the two world wars. You'd be surprised how much food you can grow in your community or your own backyard. I have about a thousand square feet of good growing space in my backyard, and every year I grow more brussels sprouts, kale, cabbage, beets, and broccoli than my kids could ever eat!

For two interesting examples of the resurgence of agriculture following economic shocks we can look to two rather unexpected places: Detroit, Michigan, and Cuba.

The Amish eschew electricity and most farm machinery, yet they farm very effectively and productively. These guys have a fair amount to teach, and their knowledge will be particularly important in times of electricity and diesel shortages. *Courtesy of Steve Hallett.*

Detroit's rise to prominence in the first half of the twentieth century was built on heavy industry, especially the automobile industry, but the city fell on hard times in a spiraling decline led by industrial consolidation, social unrest, and a "white flight" that gutted its tax base. The population of the city plummeted and nearly a third of families now live below the poverty line, with the average per capita annual income barely over $14,000. Detroit is a case study of downturn leading to tipping point and economic collapse, but some people are trying to tip Detroit back. A series of revitalization projects have significantly improved the riverfront area and some large companies have returned. Neighborhoods are improving a little and diversifying.

An encouraging aspect of the urban renewal of Detroit has been urban agriculture. Detroit's economic collapse left the economically bombed-out core of the city a food desert with anything other than the cheapest, fastest, and unhealthiest food hard to come by, but it also left an abundance of abandoned lots. The urban aggies have moved in, as have various other groups intent on developing parks and other gardens. The city has a way to go, for sure, and the response, although heartening, has not been nearly what it could have been, but Detroit is greening, and urban agriculture has played a role. Detroit can teach us a lot about how to turn a cityscape into a vegetable patch, and this is knowledge many of us will need in the coming years.

Another amazing agricultural success story is Cuba, a nation whose economy was propped up for a long time by the Soviet Union, which supplied cheap oil and bought commodity crops at elevated prices. The collapse of the Soviet Union was like a peak-oil event for Cuba, as its economy was devastated overnight. What followed was, frankly, a model of survival, community cohesiveness, and good government. It was not without its problems, for sure, but Cuba withstood the storm and reorganized much faster than anyone had expected. A major focus was urban agriculture and organic farming. Cuba converted now-useless plantations and vacant urban areas into thriving multiuse farms, and its food supply held. The period of deepest poverty was relatively short and the country rebounded into a better, more productive form of agriculture. With community spirit and leadership, much can be done.

We have enormous capacity to increase our production of healthy food in America, but farming is not quite as easy as it might seem and you can't expect to wander into a field, toss a few seeds in the dirt, and succeed as a farmer. Skill has to be acquired through experience and, ideally, through training. Many of us are working on this by trying to bring sustainable-agriculture education back into our institutions, and there are now some wonderful institutional farms around the country. The University of California, Davis, Michigan State University, and many others have developed thriving programs, and student-run farms and others are following, including a wonderful group of dedicated and hard-working students at Purdue University, where I work.[24] We need to do a much better job of training the next farmers and urban farmers of America so that we can travel as far down the necessary path as possible before the need for a major overhaul becomes a matter of survival. I couldn't ever say this as well as Ben Hartman of Clay Bottom Farm said it in a recent e-mail:

"It is often incorrectly assumed that farming is easy and that anybody could farm. The reality from our experience is that the time and energy it takes to become a successful farmer is roughly equivalent to the time and energy it takes to become a doctor—around eight years. Yet there are hundreds of institutions training the next generation of doctors while only a few offer training to students wanting to become local and sustainable farmers. That will change quickly, I think, the day a carrot costs twice as much from Walmart as it does from us."[25]

CHAPTER 13

BUILDING RESILIENT COMMUNITIES

I accordingly concluded—to my horror—that my impatience
for the establishment of democracy was itself somewhat
communistic. . . . I had wanted to move history forward in the
same way that a child pulls on a plant to make it grow faster.
I believe that you must learn to wait in the same way that you
learn to create. We have to patiently sow the seeds, water the
ground frequently where the seed is sown and give the plants
the time they need to grow. One cannot deceive a plant any
more than one can deceive history.

—Vaclav Havel, president of the Czech Republic[1]

Swimming against the Tide

It takes five years for a willing person's mind to change. Have
patience with yourself and others when treading in an area
protected by a taboo.

—Garrett Hardin[2]

The efficiency trap is a pretty disconcerting phenomenon, and I consider it to be the worst nightmare of the environmental movement. I've argued not only that our human quest for progress is threatening our future, an idea easily embraced by environmentalists, but also that the solution that appears to be the most likely to mitigate its effects—conservation through improved efficiency—is an integral part of the problem. To combat the problems we face, then, we need to swim against the tide, and not only the main tide of environmentally blind industrialists, but also the secondary current of efficiency-blind environmentalists. The views presented in *The Efficiency Trap* are decidedly not mainstream.

Human impacts can cause dramatic shifts in ecological systems, tipping them from one state to another. The deforestation of many parts of England during the Neolithic period by those pesky henge builders tipped systems from forest to moorland. Long after the cessation of hostilities, the trees remain locked out because the system had tipped. Cod are not coming back to the Grand Banks even though cod are fished much less intensively. Invasive cheat-grass in the Great Basin and purple loosestrife in the midwestern wetlands are quickly tipping those systems into an alternate steady state. Although change may seem decidedly unlikely, ecological systems can tip from one state to another quite quickly. The fire-prone forest may spend a hundred years organizing, growing, and maturing through the α, r, and K phases of its cycle and then flash through the Ω-phase, release, in a matter of days.

The Efficiency Trap is concerned with two monstrous tipping points that we face. Global climate change will bring us to serious ecological tipping points, some of which may be surprisingly rapid, and some of which are already in motion. Coral bleaching, the collapse of pine forests from bark beetle infestations, the destruction of Arctic habitat for animals such as the polar bear, and the thawing of the Arctic permafrost are not future, theoretical impacts; they are disasters that we are watching unfold. Peak oil, meanwhile, will bring us to serious economic tipping points. Our economies are under severe stress from the many limits to growth that our overblown and overconsuming societies have reached and overshot. The stagnation and failure of our energy throughput system will tip when its energy return on investment falls to a critical level, sending the global economy into a chaotic and destructive release. Systems do not always persist, and they do not change only through slow, grinding processes. They can change rapidly when they reach tipping points. These tipping points tend to be scary, but they can also be good.

Malcolm Gladwell is a journalist and author who specializes in the analysis of social and economic systems in a somewhat similar vein to the *Freakonomics* guys, Steven Levitt and Steven Dubner. Gladwell's best book, in my opinion is *Outliers*, but his first book, *The Tipping Point*,[3] was pretty good, too: *"The Tipping Point* is the biography of an idea. . . . That the best way to understand the emergence of fashion trends, the ebb and flow of crime waves . . . the transformation of unknown books into bestsellers . . . the rise of teenage smoking . . . the phenomena of word of mouth . . . is to think of them as epidemics. Ideas and products and messages and behaviors spread just like viruses do." It is a lovely exposition of the ecology of human behavior. We are, at our core, a small group animal with the ecological leanings that one would expect from a reasonably intelligent ape. Our systems can be caused to tip in various directions.

Understanding the concept of tipping points gives some hope that change can be made. When we look at our society it seems impossible that it can be changed in meaningful ways. It is stuck in a set of reinforcing cycles that lock us on destructive paths. How can farmers increase their sustainability when the system reinforces their bad habits and inhibits innovation? How can we get through to our governments when the corporations, lobbyists, and media barricade them in? How can we free ourselves from our oil addiction without abandoning our houses in the suburbs, ditching our cars, and moving out to a cabin in the Bitterroots? The system seems to be locked in this state with irresistible inertia.

Tipping points show us that systems are not always what they seem to be. We do not have to make all the changes that are required. We merely have to make enough changes to cause the system to tip, so we may not have to swim against the tide the whole way. The tide can and does shift, and we can find ourselves swimming with it. The onrushing mass of events we face is going to transform this civilization of ours at a series of unpredictable tipping points, but perhaps we can create our own tipping points, move society in a better direction, and deflect the flow of events into preferable steady states.

Again, there are some popular books that give guidance, such as *Switch*, by Chip and Dan Heath,[4] which has a nice, simple, readable message. In order to show how to switch a group through a social tipping point, Dan and Chip summon up the image of elephant riding. Getting a group of people to go in a certain direction is like getting an elephant to go where you want it. You must direct the rider, motivate the elephant, and shape the path.

The rider is the brains of the operation. The rider knows how the saddle and reins work and what the destination is, but a well-informed rider is not enough to make a large, cumbersome, stubborn animal like an elephant go the right way. The elephant is the heart of the operation, and it needs to be motivated as well as directed. With the driver properly informed and the elephant appropriately motivated, things can move. To make sure they go where you want them to, and go there as easily as possible, you shape the path. The elephant and rider will move much more effectively along a path than through the thick tangles of the jungle. If we want to move our communities into a more sustainable future, we need to do much more than just wish for it. We need to help people understand what the problems are, motivate them effectively, and put as many systems in place as we possibly can to ensure that the path of least resistance leads not to a wholesale failure, but to a better future.

The Three Pillars of Democracy

> It has been said that democracy is the worst possible form of government except all those other forms that have been tried from time to time.
>
> —Winston S. Churchill[5]

> In sum, the Court's conclusion that a constitutionally adequate recount is impractical is a prophecy the Court's own judgment will not allow to be tested. Such an untested prophecy should not decide the Presidency of the United States. I dissent.
> —Ruth Bader Ginsberg, Dissenting in *Bush v. Gore*, 2000[6]

> He who controls the present controls the past. He who controls the past controls the future.
> —George Orwell, *Nineteen Eighty-Four*[7]

If we're talking about protecting vital systems, there's a big (unmotivated) elephant in the room. I don't know if you've noticed, but our democracies are going to hell. All three of the main pillars of democracy—a government representing the people, an unbiased judiciary, and a free press—are crumbling. The US Congress has become a laughingstock. The votes of most of the US Supreme Court justices are virtually guaranteed to reflect a political and ideological bias on many issues. The most popular news media provide more opinion and propaganda than news. People are rapidly losing faith in their democracy, and with good reason.

I could go into a long political rant here, but there are plenty of good rants out there (and even more dumb ones), so I'll just state the obvious, and quickly. Our government has been hijacked by big business. The lobbying power of corporations, codified now by recent Supreme Court rulings, gives corporations much more power to elect a government than the power afforded to regular people. Many countries are deeply polarized, and few more so than the United States. It's almost impossible for most Americans to engage in a serious discussion with people of the opposite political stripe. I find it hard to listen to fervent Republicans. Few of them seem to make much sense to me, and the media has been taken over by the clowns. Put the combined wisdom and honesty of Rush Limbaugh, Bill O'Reilly, Sean Hannity, and a dozen others on the back of a postcard with a thick marker pen and you'll still have plenty of room for the address and stamp. Decent, honest reporting does not seem to be a money maker, and real reporting is increasingly drowned out by news with an agenda. Well, I'm sorry, but if it has an agenda, it's not news. There's this guy called Rupert Murdoch . . . oh, never mind.

What on earth can we do to protect democracies against this onslaught? I must admit that I don't think most people can do much to influence those at the top of government hierarchies. Bugging your representatives remains essential, as does voting in elections, and the most democratic thing you can do is to speak truth to power, but I think local activities will be much more productive and useful. Work locally to direct the driver, motivate the elephant, and shape the path.

National and state governments, dominated by wealth and influence, are out of the reach of most people. It's hard to break into that circle even if you want to, and it's not clear that your voice is heard inside it. Your local government, however, is a different matter. Show up at council meetings or join the council or school board. Support your local NPR station, start a newspaper, give talks. Much can be done. Since you are now going to the local council meetings, let everyone know what the council is planning. Start a transition initiative or a local garden, or lobby for the extension or creation of a bike path. Object to any rezoning that puts concrete over good soil. Question the benefits of tax breaks that bring businesses into town. Start a food co-op. Examine the council's energy plan and create an energy-descent plan. Few of your leaders are able to participate in the needed transition, so you're going to have to make changes yourself.

The Three Pillars of Sustainability

> The global economy is now so large that society can no
> longer safely pretend it operates within a limitless ecosystem.
> Developing an economy that can be sustained within the finite
> biosphere requires new ways of thinking.
>
> —Herman Daly, "Economics in a Full World"[8]

It is useful to think of sustainability as having three components: an environmental component, an economic component, and a social component. It is awfully difficult to sustain a system without also sustaining the systems upon which it is dependent. It is hard to protect the environment, for example, if economic conditions are not favorable. The midwestern farming landscape would be much more environmentally sustainable if some of our huge acreages of corn and soybeans could be cut up into medium-sized, mixed-crop-and-livestock family farms, but farmers need to be able to make this transition without going broke in the process. Converting buildings from fossil fuels to alternative energy may enable them to function in the future, but can you afford to do it today?

The social component is also vital. Economic growth can bypass large sectors of society. Our high-throughput systems have delivered huge improvements for the standard of living of many people around the world, but they have been distributed very poorly and have brought a number of other problems. A society that supports a wealthy few at the expense of the larger population is unlikely to persist. Social problems weaken us all, and without a cohesive society neither environmental nor economic sustainability is possible. The economic, social, and environmental components of sustainability are inextricably linked; transitioning smoothly from our current model to a future one will require progress in all three areas.

One of the reasons we struggle so much to approach sustainability is because we prioritize its three pillars in the wrong order. We treat the economy as the overarching goal. If the economy is not growing, we pull out all the stops to get it restarted, society and environment be damned. The second emphasis is on society. We can fight over who gets the spoils so long as there is economic growth. Should we invest in the wealthy so that they can make the economy turn over even faster, trickling wealth down to the rest of us, or should we distribute more wealth directly to the poor and middle class to make sure everyone is driving the economy?

Environmental protection is an afterthought. We can't afford to clean up the environment, it is claimed, when the economy is stagnant. If the economy is bubbling along, we can probably spare some of the money for the environment, even though most people think they should have first dibs on the cash. The economy and the environment are viewed by many as being in competition.

Environmental, social, and economic sustainability are linked and inter-dependent, to be sure, and none can thrive when the others are neglected, but there is a definite hierarchy, and the environment is master. Societies and economies are utterly dependent upon it. We can power economies and build societies on the back of the environment for only so long. If we do not sustain our environment—and our civilization never has—we will eventually fail.

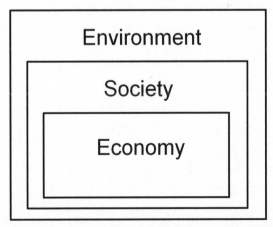

The economy is dependent on society, which is dependent on the environment. No system can be sustainable unless the systems upon which it is dependent are also sustained. *Courtesy of Steve Hallett.*

The Three Pillars of Resilience

> If trouble comes when you least expect it then maybe the thing
> to do is always expect it.
>
> —Cormac McCarthy, *The Road*[9]

The long-term goal of society should be sustainability, and perhaps, one day, we will find a way to live within the limits of a protected sustainable environment, but this is not possible now. Our society has overshot its capacity to sustain itself, so today's goal cannot be sustainability. Instead, it must be resilience. While sustainability is the capacity to remain in a steady state, resilience is the capacity to recover from shocks. Nothing can sustain a system that is in overshoot and nothing can now prevent the shocks from coming. Resilience is a special kind of strength. It is not the strength to hold back the tide, but the strength to bend without breaking as the tide inevitably rushes in.

The three pillars of resilience are diverse and appropriately scaled energy systems, diverse, low input food systems, and decentralized networks. We repeatedly hear calls for more energy and food, and for greater efficiency in their production, but more and bigger is not what we need. We need systems that won't fail when the pressure is on and are nimble enough to adjust quickly. Decentralization will prevent shocks from spreading throughout the system, like firebreaks in the forest. Diversity offers the system options from which it can choose as the pressures upon it shift and change.

We begin at home, where the motto is self-sufficiency. Reduce your energy needs so that they match your energy production. If you can be energy independent, you have a huge advantage. Also consider your food and water independence: plant a victory garden; collect rainwater.

Self-sufficiency is unlikely to be as successful as it may seem, however, and you will also need to lean on the people around you, so it makes sense to do what you can to prepare your community. Can your community weather the coming storm if it can no longer rely on the region or nation to supply the resources and services it needs?

The energy problem is something that some communities can tackle, and some have tackled it already, but it will be virtually impossible for others. A small, compact, well-connected, relatively wealthy rural community (a rare thing) may be able to organize an effective community energy system. A poor, inner-city district will have basically no chance of doing this.

Similar problems exist with food and water, although there are generally more options for local food production than you might think. Remember Ben and Rachel, at Clay Bottom Farm, who produce food for a community on five acres, and Carl Tricamo, who produces food on his front lawn. You can feed a family on a quarter-acre lot if it comes to it. Community gardens on vacant lots and local farms may have a huge role to play as it becomes harder and more expensive to move food produced in rural areas to the cities and suburbs.

The most important role of communities might be in the area of security. I don't think the solution to security lies at the level of the individual home. Sure, get a gun if that's your thing, but security is a bigger issue than the height of the walls around your compound. And the security of communities is something that we can improve quickly. Do you even know most of the people who live around you? If you live in a small, rural community, you probably do, and that brings strength, but if you live in a featureless bedroom community I'll bet you don't, and that's a problem. Just start in this simple way: say "hi" to your neighbors.

A big problem with our communities is the way in which we have become disconnected, and many of our population centers are not communities at all. We get around this today because these centers are connected to a larger, central hub. The suburbs have not needed even the slightest vestige of self-sufficiency because suburbanites are within easy reach of all the services they need by car. The centralized model is deeply vulnerable, and the suburbs will need to reorganize into more resilient hubs of their own if they are to avoid the worst of the coming collapse. This reorganization of our networks across the landscape will be essential when transportation becomes less effective and more expensive. An increasing number of the goods and services that you depend upon will have to be generated locally as the daily commute to central hubs declines. So what is there locally? Are there opportunities for employment? Are there shops that you can get to easily? What about hospitals?

Our resilience would be made much stronger if we could decentralize our society and better enable communities to function locally. Europeans will generally find this transition easier than Americans because they retain a much more modular structure, both at the landscape level and in the cities. While America is an efficient construct of the petroleum interval, Europe is an upgrade of systems that were laid down in the centuries before the Industrial Revolution. European farming landscapes tend to be much more diverse, and European cities have relatively diverse cores that require less repurposing. Residential buildings, cafes, and small retailers live cheek by jowl throughout the city. American cities, built largely through the petroleum interval, have larger expanses of suburbs and a much greater reliance on big-box stores and long-distance supply chains. The centralized system is very efficient now, but it will be much harder to reorganize it into interconnected networks that will function in the future. Our preparations need to begin with communities.

There is a long list of things that our national and state leaders should be doing to prepare, but few understand the nature of the problem. Many policies need to be modified; yet more require a complete U-turn. We should stop subsidizing the fossil-fuel industries and instead provide subsidies for local and community energy projects. We should stop subsidizing farmers to produce commodity feedstocks and instead subsidize them to produce food. Zoning committees should stop allowing houses and businesses to be planted on farmland. Regulations that inhibit home, local, and urban food production should be removed. Road construction should virtually cease. Light, electric, local rail systems should be rebuilt. Environmental protection laws should be strengthened and enforced.

The changes we need to make to laws and regulations, the changes we need to make in the collection and disbursement of funds, and the changes we need to make regarding the types of infrastructure projects we need to pursue represent impossible departures for governments, and most of these changes are obviously not going to happen. What I'm proposing here, in essence, is for governments to peel back support from large businesses, especially extractive industries. Instead, governments should support small businesses. They should inhibit oversized energy and agriculture systems and promote smaller, localized ones. They should support the ability of their countries and communities and citizens to decentralize and organize new networks for a post-peak world. Where government is unwilling or unable to make such changes—and that will be in most places—it will fall to individuals and communities to make their own preparations.

We face enormous, perhaps existential, challenges to our civilization, but we have been assured by the majority of our politicians, captains of industry, engineers, economists, and scientists that solutions abound. Our engineers regale us with prototypes for technological advances that are just around the corner, and our economists assure us that advances not yet imagined will spring to the fore when the markets demand them. Part of me still wants to hope they are right, but calming words and reassurances are the springs and clamps of the efficiency trap—and they sound more and more desperate every year.

New technologies will not liberate us. Even if they do materialize they will not reduce our consumption of resources. They will not reduce our demand for energy. They will not diminish our impacts on the climate. And they will not bring about a better world. Rather, they will do what new technologies have always done: they will spur economic growth and increase our destruction of the natural world. Our politicians, meanwhile, exhort us to save energy, save water, and save fuel by being more efficient. Together, they tell us, we can save the world. If only it were that simple.

No unexpected technological leap is going to spring up and save us, but it's much more than that. We still refuse to answer the more important questions: save us from what, exactly, for how long, and at what cost? Perhaps Aldous Huxley had it right when he said, "Technological progress has merely provided us with more efficient means for going backwards."[10]

We can't continue this maniacal growth, that much is sure, but perhaps we can find some way to arrest the madness and settle down into a steady-state economy that sustains itself and limits our impacts on the planet. It seems like a worthy goal, and what could be a better goal? This, too, is unattainable. It is too late to slow this runaway train down before it tumbles into the chasm. The train is not only moving too fast, but it is still accelerating. Even worse, the train is already on the broken bridge. Our global economy does not have much time left before today's periodic downturns become increasingly serious, reinforcing failures.

It's time to realign our sights. We must abandon not only growth, but also the vain hope that we can sustain the society we have built. Efficiency does not deliver on its promise. At best, it allows us to strike a Faustian bargain. Neither growth nor sustainability is an option anymore. We need to build our resilience. It's time to batten down the hatches and prepare for the coming storm.

EPILOGUE
THE KEY THAT UNLOCKS THE EFFICIENCY TRAP

Efficiency sets two pervasive and counterintuitive traps. The first trap is to convince us that we have found a way of conserving without abandoning progress and growth. We earnestly adopt a more efficient machine that can do the same work with less energy and save us time or money. But more efficient technologies drive progress, and the time and money we have saved is soon used for more consumption. Perhaps we use the more efficient machine more, or for additional tasks, or perhaps we spend the saved time and money consuming something else. Efficiency and conservation are not the same things. Efficiency promises to enhance conservation, but it actually drives consumption.

This first efficiency trap is relatively easy to understand, but it can be tantalizingly difficult to unlock. Most people I talk to tend to think that they are in control of this particular trap. The new Prius owner claims that she doesn't drive more just because she has a more fuel-efficient car: well, maybe just a tad more. The neighbor who recently added new insulation to his house assures me that he keeps the thermostat at the same temperature as before: well, maybe just a tad higher.

Unlocking this first efficiency trap requires us to use efficiency measures strictly as conservation measures, and this is more difficult than it would seem: but it can be done. One of the best examples of a potentially valuable efficiency improvement is in home heating. Highly efficient furnaces have had the effect of enabling us to build bigger, flimsier houses because the costs of wasted energy are no longer prohibitive. This is a classic efficiency trap, and we consume prodigious amounts of energy in huge, poorly insulated houses as a result. There are those, however, who, despite the fact that they may own a small, well-insulated home, still choose the most efficient heating system. This is conservation. The key, of course, is to think not of the proportion of energy that will be saved by an efficient technology but of the total amount of energy that will be used. What's even more important than the energy saved, however, is the fact that it may still be possible to find a way to heat the home when fuel prices rise.

The second efficiency trap is much more difficult to see. This trap convinces us that efficiency will enable us to better control and strengthen complex systems. In most cases, alas, the very contrary is the truth.

Cellular systems, ecosystems, and civilizations are, at base, thermodynamic systems. Energy flows through systems in cascades, from layer to layer, driving all their transformations from one state to another: sunlight to plant, plant to animal, and animal to microbe, for example, or coal to steam, steam to turbine, and turbine to electricity. The system uses energy, eventually releasing it as useless heat. In the requirement for an energy input and the establishment of an energy cascade, all systems are the same, but there is one vital difference between natural systems and human systems: natural systems receive a fixed energy input that never varies.[1] If we could set a strict limit on energy use, we could build a human system that is just as stable as an ecosystem.

This would seem to be a prescription for the abandonment of progress, but it is not. Fixed energy limits have not limited the progress of nature. And they need not limit the progress of society. The fixed energy limits of the biosphere have not curtailed progress. They have done the exact opposite. They have created balanced, sustained, competitive systems in which responsiveness to the environment is essential not only through long-term, renewing collapses, but also through short-term, regulating feedbacks. Fixed limits have spurred the radiation of life into increasingly complex niches and led to its flourishing diversity. Life has blossomed into the myriad forms we see, each life-form finding a place in a great energy cascade. Life has—quite unwittingly—evolved the most stunningly efficient systems, and so here is the great irony. Efficiency can be a devastating trap, but a sustained energy supply not only unlocks the efficiency trap; it also unleashes efficiency to build not peril and overshoot, but diversity and beauty.

We seem to be condemned to crash and burn after each flurry of growth and development, and our modern world seems to be following history's familiar path. We harness energy and resources from the environment in increasing volumes, with ever-increasing efficiency, and we apply them to build a society that becomes bigger, more complex, more extractive, and more polluting. As we struggle to keep growing into a shrinking future we redouble our efforts to increase our efficiency, but in vain. The trend lines meet and then cross. Society's growing demands exceed the environment's ability to supply, and we collapse. But through the millennia, as we have continually experimented with systems of civilization, repeating the same, tired old playbook of rise and fall, systems that display all the secrets of sustainability have endured alongside us. We have, in fact, lived within these systems, as a part of them, and we've used them to power our own—yet we still haven't noticed.

A bristlecone pine in the Colorado Rockies was recently dated at five thousand years old. Its seed germinated when the first classic civilizations were being established along the great rivers of China, India, the Middle East, and North Africa. The ancient Britons were building Stonehenge. The horse had not yet been domesticated in Eurasia, although it had recently been sent into extinction in North America. And this is a single tree: just one individual on a shifting, moving, adapting landscape of ecosystems that have never failed.

Human systems resemble natural systems in many ways, but they have never emulated natural systems in the most important regard: they have never solved the sustainability puzzle. Human systems always bring about their own demise while natural systems never do. Natural systems are swatted aside by continental drift and glaciations, to be sure, but they do not destroy themselves. They hold the key that unlocks the efficiency trap. They are able to adapt to the changing times without failing, and they do so in two ways. First, they can swap out their complement of species from a diverse collection, and second, the species themselves evolve.

The huge ice sheets of the last ice age finally left the Midwest about fourteen thousand years ago. Ice a mile thick covered the landscape as far south as Indianapolis. The ecosystems to the south of the ice sheets were similar to today's Arctic tundra, and south of those were the boreal forests. As the ice retreated and the earth warmed, the boreal forests marched back north.

The disruptions were huge, like the ones we face from anthropogenic global climate change today (though today's changes are happening much more quickly), but the ecosystems moved with the times. The ecosystems of Indiana were transformed through tundra, then boreal forest, eventually becoming the present mosaic of deciduous forest, prairie, and wetland (and there is a tiny remnant of the old boreal forest in the nooks and crannies of Turkey Run State Park).[2] The ability of ecosystems to adapt is phenomenal. They can dance across the landscape, roiling with change, and yet maintain their essential structure and function.

At the level of species, the ability to evolve is the very definition of life. We are, all of us, just the latest combinations of chunks of stardust that found a way to replicate with modifications billions of years ago. What were once the simplest of cells have become complex organic machines, but we still play out the same game of thermodynamics, and we still tell the story of life in the oldest of all languages: DNA.

The resilience of ecosystems, then, comes from their ability to adapt, and this, in turn, operates by enhancing the abundant diversity from which they can select. The diversity of species comes from evolution by natural selection, and the key to evolution is sex. To find our steady state and solve the sustainability puzzle, we need to abandon the relentless quest for dominance. We need to abandon our visions of progress as growth, because progress measured as growth is transient. Only progress in diversity, equality, and beauty can stand the test of time. We need to learn to live within limits. So relax, take it easy, spend more time with the one you love, and remember: the key that unlocks the efficiency trap is sex.

The efficiency trap cannot, alas, be unlocked retroactively. The society in which we live must, in the coming decades, face the wrath of nature's long-term feedbacks swinging into devastating force. We have failed the test again. Our society has outgrown its energy supplies and resource base, degrading its environment. We have, once again, failed to solve the sustainability puzzle—but at least we now know how to solve it. Perhaps a new civilization, emerging from the ashes of ours, will finally unlock the efficiency trap.

ACKNOWLEDGMENTS

Many thanks to all the people who helped me formulate these concepts and bring them together, first in my head and then in *The Efficiency Trap*, including those who debated and argued with me, patiently explained things to me, edited my verbose prose, and corrected a thousand errors. I am very fortunate to have a wonderful, patient, and very smart family, and a group of highly intelligent friends and colleagues who tolerate my ramblings. To Shelley, Dan, Pat, Sami, and Lindsey: How on earth do you put up with me? I love you, and I'm glad you do. When I write, I am always thinking of the family that is either back home in England or no longer with us. It makes me happy to wonder what you would say, and it helps me to focus. So thanks to Mum, Dad, John, Chris, Uncle Deryck, and Uncle John. Special thanks to the people who waded through early drafts, pointed out things that were clumsy or dumb, and put in the long hours of coffee- or beer-drinking with me to help me improve the book: Kevin Gibson, Tamara Benjamin, Jody Tishmack, and Jeff Dukes. Thanks also to the students at Purdue University who are forced to listen to me, especially those in my honors class and at the student farm. Lots of my ideas have been formulated and refined in my interactions with you. Thanks to the growers who have shared their vision and wisdom: Adam Moody, Kevin Cooley, Neil Moseley, Ben Hartman, and Ian Thompson. Special thanks to my boss, Peter Goldsbrough, for being so supportive, despite my unconventional ways. Thanks to Ashley Holmes and David Johnson for the use of photographs. Many thanks to the team at Prometheus Books for bringing the book together so beautifully.

NOTES

Prologue: *The Eighth Deadly Sin*

1. E. Luttwak, *Turbo-Capitalism: Winners and Losers in the Global Economy* (New York: Harper Perennial, 2000). Luttwak is an interesting guy, a longtime military strategist, and a consultant to many organizations within the US government (for example, the State Department and the National Security Administration) and to other countries. One of his themes is that efficiency can be a disaster in military strategy, making you too predictable. He is also a rather off-the-wall commentator on various other topics, sometimes drawing strong criticism.

2. Well, to be honest, that would work for me . . . cram the whole opera into five minutes, I say.

3. GDP is a horrible measure. It does not distinguish between activities that are good and those that are bad. Building something generates GDP, but so does tearing it down, and so does the waste management industry when it discards it; and so does the removal of the natural resource to put it in the marketplace in the first place. All activities contribute to GDP, whether they are progressive or not. It also misses lots of things. Volunteer work may be of great community service, but it does not score in terms of GDP. So GDP ignores (and even views as economic gain) activities that deplete natural resources, but it does not value activities that support community. The reliance on GDP as a measure of growth also sends indicators that GDP is a measure of progress, but progress toward what?

Part 1: THE EFFICIENCY PARADOX

Chapter 1. *The Petroleum Interval*

1. J. M. Diamond, *Collapse: How Societies Choose to Fail or Succeed* (New York: Penguin, 2005); T. Homer-Dixon, *The Upside of Down: Catastrophe, Creativity, and the Renewal of Civilization* (Washington, DC: Island Press, 2006); S. G. Hallett and J. Wright, *Life without Oil: Why We Must Shift to a New Energy Future* (Amherst, NY: Prometheus Books, 2011); J. A. Tainter, *The Collapse of Complex Societies* (Cambridge: Cambridge University Press, 1988).

2. Diamond, *Collapse*; Homer-Dixon, *Upside of Down*; Hallett and Wright, *Life without Oil*; Tainter, *Collapse of Complex Societies*. A colleague of mine at Purdue University, Nick Rauh,

has done some very cool work in Anatolia, Turkey, where he and his team have shown that the decline of some of the great Mediterranean cedar forests coincides with the timeline of the fall of Rome. U. Akkemik, H. Caner, G. A. Conyers, M. J. Dillon, N. Karlioglu, N. K. Rauh, and L. O. Theller, "The Archaeology of Deforestation in South Coastal Turkey," *International Journal of Sustainable Development & World Ecology* 19, no. 5 (October 2012): 395–405.

3. W. S. Jevons, *The Coal Question: An Inquiry Concerning the Progress of the Nation and the Probable Exhaustion of Our Coal-Mines*, 3rd ed., ed. A. W. Flux (1865; repr., London: Macmillan, 1906).

4. "Disposal" is an unfortunate word choice here, but of course it's the right one.

5. Hallett and Wright, *Life without Oil*.

6. "It was a field in area. . . . I provided it with six decks . . . three times three thousand six hundred [units unclear] of raw bitumen I poured into the bitumen kiln . . . and I set to the oiling. . . . I went into the boat and sealed the opening. . . . Six days and seven nights came the wind and flood." *The Epic of Gilgamesh*, introduction and translation by M. G. Kovacs (Palo Alto, CA: Stanford University Press, 1985). In the Bible, God commands that Noah build an ark of gopher wood and that he, "pitch it within and without with pitch." The Ancient Mesopotamians also used oil from seeps for lighting and other things.

7. K. S. Deffeyes, *Beyond Oil: The View from Hubbert's Peak* (New York: Hill and Wang, 2005); M. K. Hubbert, "Nuclear Energy and the Fossil Fuels," spring meeting of the American Petroleum Institute, San Antonio, TX, March 7–9, 1956; M. R. Simmons, *Twilight in the Desert: The Coming Saudi Oil Shock and the World Economy* (New York: John Wiley & Sons, 2006); C. J. Campbell and J. H. Laherrère, "The End of Cheap Oil?" *Scientific American*, March 1998, pp. 78–83; R. Heinberg, *The Party's Over: Oil, War, and the Fate of Industrial Societies* (Gabriola Island, CA: New Society, 2003); J. K. Leggett, *The Empty Tank: Oil, Gas, Hot Air, and the Coming Global Financial Catastrophe* (New York: Random House, 2005); J. H. Kunstler, *The Long Emergency: Surviving the End of Oil, Climate Change, and Other Converging Catastrophes of the Twenty-First Century* (New York: Grove Press, 2005).

8. R. Heinberg, *Blackout: Coal, Climate, and the Last Energy Crisis* (Gabriola Island, CA: New Society, 2009).

9. R. Heinberg and D. Fridley, "The End of Cheap Coal," *Nature* 468 (2010): 367–69; T. Patzek and G. Croft, "A Global Coal Production Forecast with Multi-Hubbert Cycle Analysis," *Energy* 35 (2010): 3109–22.

10. Heinberg, *Blackout*.

Chapter 2. The Conventional Wisdom

1. R. Heinberg, *The End of Growth: Adapting to Our New Economic Reality* (Gabriola Island, CA: New Society, 2011).

2. D. Capiello and T. Krisher, "Obama Announces New Fuel Efficiency Standards, *Huffington*

Post, http://www.huffingtonpost.com/2011/07/29/obama-fuel-economy-deal_n_913341.html (accessed August 3, 2011).

3. P. Krugman, *End This Depression Now* (New York: W. W. Norton, 2012). I need to make a couple of comments here. I admire Paul Krugman, but the problem with Democrats is that they actually have good ideas how to do this . . . which is dangerous. The Republicans are a little safer because, while they claim they want to grow the economy, they are actually clueless.

4. That's a little generous. Relatively little constructive has been done in the last decade or so, as our politics have increasingly descended into the realms of the pointless and peripheral.

5. This is quite a mixed bag. The monetary costs of some things are still low or falling but the real costs are rising. The quality of many goods, such as wood, is also falling.

6. R. U. Ayres and E. Ayres, *Crossing the Energy Divide: Moving from Fossil Fuel Dependence to a Clean-Energy Future* (Upper Saddle River, NJ: Pearson Prentice-Hall, 2012).

7. Duke Energy, "Who We Are," http://www.duke-energy.com/environment/who-we-are .asp (accessed April 2, 2012)

8. B. McKibben, *Deep Economy: The Wealth of Communities and the Durable Future* (New York: Macmillan, 2008); L. R. Brown, *Plan B 4.0: Mobilizing to Save Civilization* (New York: W. W. Norton, 2009); R. Heinberg, *The Party's Over: Oil, War, and the Fate of Industrial Societies* (Gabriola Island, CA: New Society, 2003).

9. The American Council for an Energy Efficient Economy (ACEEE) has a number of documents on its website (http://www.aceee.org/), including "Reducing the Cost of Addressing Climate Change through Energy Efficiency: Consensus Recommendations for Future Federal Climate Legislation in 2009," a consensus report supported by a range of other groups, including the Sierra Club, the Alliance to Save Energy, and the National Association of Energy Service Companies (http://aceee.org/files/pdf/ReducingtheCostofAddressingClimateChange .pdf [accessed January 8, 2013]); K. Ehrhardt-Martinez and J. A. Laitner, "The Size of the US Energy Efficiency Market: Generating a More Complete Picture," Report E083, American Council for an Energy-Efficient Economy (Washington, DC: ACEEE, 2008); Armory B. Lovins, "The Negawatt Revolution," *Across the Board* 27 (1990): 18–23.

10. *Food, Inc.*, directed by Robert Kenner (River Road Entertainment, Magnolia Pictures, Participant Media, 2008).

11. Crikey. It seems like only yesterday that I was telling my classes "six billion."

12. More on this later, but where on earth do they come up with these numbers? I can't believe how many times I've heard this "fact" that the global population will reach about nine billion by midcentury and then (because we'll all be nice and healthy, wealthy, and wise at that time?) will remain stable at that number.

13. More on the Green Revolution in chapter 6.

14. A gene from *Bacillus thuringiensis* can be expressed in plants, enabling the plants to produce a toxin that kills insect pests. A mutant version of the gene attacked by the herbicide glyphosate was found in a bacterium unaffected by glyphosate, and it has been put into

plants so the herbicide, which kills or damages most weeds (until they evolve resistance), can be sprayed.

15. The concept of the carbon footprint was introduced more broadly as the concept of the ecological footprint. W. E. Rees and M. Wackernagel, "Ecological Footprints and Approached Carrying Capacity: Measuring the Natural Capital Requirements of the Human Ecosystem," in *Investing in Natural Capital*, ed. A. M. Jansson, C. Folke, and R. Costanza (Washington, DC: Island Press, 1994).

Chapter 3. The Unconventional Wisdom

1. M. Twain. *Mark Twain's Notebook*, ed. A. B. Paine (New York: Harper and Brothers, 1935).

2. W. S. Jevons, *The Coal Question: An Inquiry Concerning the Progress of the Nation and the Probable Exhaustion of Our Coal-Mines*, 3rd ed., ed. A. W. Flux (1865; repr., London: Macmillan, 1906). Here is the more complete quote:

> It is very commonly urged, that the failing supply of coal will be met by new modes of using it efficiently and economically . . . [that] the coal thus saved would be, for the most part, laid up for the use of posterity. It is wholly a confusion of ideas to suppose that the economical use of fuel is equivalent to a diminished consumption. The very contrary is the truth. As a rule, new modes of economy will lead to an increase of consumption according to principles recognized in parallel instances. The economy of labour effected by the introduction of new machinery throws labourers out of employment for the moment. But such is the increased demand for the cheapened products, that eventually the share of employment is greatly widened. . . . So material nature presents to us the aspect of one continuous waste of energy and matter beyond our control. . . . The Dowlais Ironworks would require no less than a thousand large windmills! Of course it is useless to think of substituting any other kind of fuel for coal. We cannot revert to timber fuel, for nearly the entire surface of our island would be required to grow timber sufficient for the consumption of the iron manufacture alone. . . . Its natural supply is far more limited and uncertain than that of coal, and an artificial supply can only be had by distillation of some kind of coal at considerable cost. To extend the use of petroleum, then, is only a new way of pushing the consumption of coal, and while the sun annually showers down upon us about a thousand times as much heat-power as is contained in all the coal we raise annually; yet the thousandth part, being under perfect control, is a sufficient basis for all our economy and progress.

3. OK, so plenty of other things happened . . . the small matter of two world wars, for example.

4. Apologies for the units here, as I've simplified a little for access. Jevons expressed this data as the number of pounds of water raised a foot by eighty-four pounds of coal and the number of pounds of coal per horsepower per hour. Not all his data seem to be fully cited. I have converted the data into slightly simpler units, but the bottom line is that it gives an idea of how much coal was needed to pump water out of mines. Bear in mind that mines were generally a fair bit deeper than one foot.

	1*	2*	
1769	30	6	(Atmospheric engine: Thomas Savery, 1700; Thomas Newcomen, 1710)
1772	18	10	(Low pressure engine: James Watt, 1768)
1776		22	
1788		27	
1820		28	(High pressure engine: Oliver Evans, 1804)
1825	10		(Double cylinder engine: Arthur Woolf, 1804)
1830		43	
1850	6		
1859		80	
1875	3		Compound marine engine: John Elder, 1850–1891)
1900	1		
1* lbs coal/horse power/h			
2* lbs water raised 1 ft by 84 lbs coal (x1,000,000)			

5. He went to night school and learned to read and write as a young adult.

6. N. Rosenberg, "Energy Efficient Technologies: Past, Present, and Future Perspectives," conference at Oak Ridge National Laboratories, Oak Ridge, TN, August 1989, cited in H. Inhaber, *Why Energy Conservation Fails* (Westport, CT: Quorum Books, 1997).

7. L. A. Brookes, "A Low-Energy Strategy for the UK by G. Leach: A Review and Reply," *Atom* 269 (1979): 3–8; J. D. Khazzoom, "Economic Implications of Mandated Efficiency Standards for Household Appliances," *Energy Journal* 1 (1980): 21–39; J. D. Khazzoom, "Energy Saving Resulting from the Adoption of More Efficient Appliances," *Energy Journal* 8 (1987): 85–89. B. Alcott, "The Jevons Paradox," *Ecological Economics* 54 (2005): 9–21; H. Herring, "Does Energy Efficiency Save Energy? The Debate and Its Consequences," *Applied Energy* 63 (1999): 209–26; J. M. Polimeni, K. Mayumi, M. Giampetro, and B. Alcott, *The Jevons Paradox and the Myth of Resource Efficiency Improvements* (London: Earthscan, 2008); H. D. Saunders, "The Khazzoom-Brookes Postulate and Neoclassical Growth," *Energy Journal*, October 1, 1992.

8. P. Newman, "Greenhouse, Oil, and Cities," *Futures* 5 (1991): 335–48; J. Cherfas, "Skeptics and Visionaries Examine Energy Saving," *Science* 251 (1991): 154–56.

9. H. Herring and S. Sorrell, ed. *Energy Efficiency and Sustainable Consumption: The Rebound Effect* (Basingstoke, UK: Palgrave MacMillan, 2009). This book is an excellent resource for data on the rebound effect. It is well referenced, and a number of chapters are particularly good.

Steve Sorrell gives a great overview of available data on rebound in chapter 2, Robert Ayres and Benjamin Warr point out the intimate link between efficiency and the development of civilization in chapter 6, and Roger Levett gives a fascinating assessment of the policy implications of rebound in chapter 9.

10. S. Sorrell, "The Evidence for Direct Rebound Effects," chapter 2 in Herring and Sorrell, *Energy Efficiency and Sustainable Consumption*. D. L. Greene, J. R. Khan, and R. C. Gibson, "Fuel Economy Rebound Effect for US Household Vehicles," *Energy Journal* 20 (1999): 1–31; K. A. Small and K. van Dender, "A Study to Evaluate the Effect of Reduced Greenhouse Gas Emissions on Vehicle Miles Traveled," Report 02-336 to the California EPA and California Energy Commission, 2005; O. Johansson and L. Schipper, "Measuring Long-Run Automobile Fuel Demand: Separate Estimations of Vehicle Stock, Mean Fuel Intensity, and Mean Annual Driving Distance," *Journal of Transportation Economics and Policy* 31 (1997): 277–92; J. Houghton and S. Sarkar, "Gasoline Tax as a Corrective Tax: Estimates for the United States 1970–1991," *Energy Journal* 17 (1996): 103–26.

11. Sorrell, "Evidence for Direct Rebound Effects"; A. Wilson and J. Boehland, "Small Is Beautiful: US House Size, Resource Use, and the Environment," *Journal of Industrial Ecology* 9 (2005): 277–87.

12. H. Inhaber, *Why Energy Conservation Fails* (Westport, CT: Quorum Books, 1997).

13. T. Jackson, *Prosperity without Growth: Economics for a Finite Planet* (London: Earthscan, 2012).

14. R. U. Ayres, L. W. Ayres, and B. Warr, "Exergy, Power, and Work in the US Economy 1900–1998," *Energy* 28 (2003): 219–73.

15. And this is just the efficiency with which it is delivered to the consumer. The consumer then suffers further loss as it is put to use.

16. And you really wish you could have the SUV back because the roads are pothole hell now, to boot.

Part 2: EFFICIENCY TRAPS

1. A. Price, *Slow-Tech: Manifesto for an Over-Wound World* (London: Atlantic Books, 2009).

2. Anstey Museum, Anstey, Leicestershire, UK, as reported in *Guide to Rural England: Leicestershire*, 4th ed., a travel guide available at http://www.scribd.com/doc/59935407/Guide -to-Rural-England-Leicestershire-Rutland (accessed June 4, 2012).

3. Yep, it's glisters, not glitters, from William Shakespeare's *The Merchant of Venice*.

4. U. Sinclair, *The Jungle* (New York: Doubleday, Jabber, 1906).

5. H. Ford and S. Crowther, *My Life and Work* (New York: Doubleday, Page, 1922).

6. J. Steinbeck, *Cannery Row* (New York: Viking, 1945).

7. F. W. Taylor, *The Principles of Scientific Management* (New York: Harper & Bros., 1911).

Chapter 4. Energy Efficiency Traps

1. *The End of Suburbia: Oil Depletion and the Collapse of the American Dream*, directed by Gregory Greene (Electric Wallpaper Company, Ontario, Canada, 2004). Kunstler's most influential nonfiction books that address similar themes are J. H. Kunstler, *The Long Emergency: Surviving the End of Oil, Climate Changes, and Other Converging Catastrophes of the Twenty-First Century* (New York: Grove Press, 2005), and J. H. Kunstler, *The Geography of Nowhere: The Rise and Decline of America's Man-Made Landscape* (New York: Simon & Schuster, 1994).

2. D. Superville, "Obama Announces Fuel Standards for New Vehicles," *Huffington Post*, http://www.huffingtonpost.com/2011/08/09/obama-fuel-standards-big-vehicles_n_922482.html. (accessed June 6, 2012).

3. There are no such reductions, of course, and no reductions are actually required, only efficiency improvements with the reductions calculated ignoring rebound.

4. Corporate Average Fuel Economy, which is the average fuel efficiency across a company's fleet.

5. I have grown a little tired of the constant use of the word *foreign* in that phrase.

6. Sorry, again, Toyota execs. Now I'm comparing your superefficient car to a jet liner.

7. Sometimes as few as 550 passengers—and the fuel efficiency calculation made assumes the A380 is full and the Prius is carrying only the driver.

8. "History of ENERGY STAR," ENERGY STAR, US Environmental Protection Agency, http://www.energystar.gov/index.cfm?c=about.ab_history (accessed July 7, 2012).

9. Having to wing it is not much of a novelty for me in class, I'm sorry to say. I'm not always the best-prepared professor, but I find we often learn a lot more when we go off on tangents (as was the case here), although my students don't always seem to appreciate it.

10. (1) None. They're all too busy trying to design the perfect one. (2) Three. One to hold the ladder, one to handle the lightbulb, and one to translate the instructions from Chinese to English.

11. LED (light-emitting diode).

12. Paul Rubens, "LED Bulbs: The End of the Lightbulb as We Know It," BBC In Depth, 2012, http://www.bbc.com/future/story/20120314-the-end-of-the-lightbulb (accessed March 14, 2012).

13. R. Fouquet and P. Pearson, "Seven Centuries of Energy Services: The Price and Use of Light in the United Kingdom (1300–2000)," *Energy Journal* 27 (2006): 139–77.

14. Available from the NASA website at http://www.earthobservatory.nasa.gov/IOTD/view.php?id=896 (accessed January 8, 2013). This is not only a pretty map, it is also an extremely interesting and instructional one. It also shows the location of fires in red and natural gas flares in yellow. The ecological footprint of human impacts on the planet is particularly obvious when the sun goes down.

15. Tianjin Economic-Technological Development Area homepage, http://www.teda.gov.cn/html (accessed July 4, 2012).

16. US Green Building Council, "About USGB," http://www.usgbc.org/DisplayPage. aspx?CMSPageID=124 (accessed July 12, 2012).

17. Efficiency Cities Network home page, http://www.efficiencycities.org/ (accessed July 7, 2012).

18. Energy Sector Management Assistance Program, "Energy Efficient Cities Initiative (EECI)," http://www.esmap.org/esmap/EECI (accessed July 7, 2012).

19. G. West, "The Surprising Math of Cities and Corporations," TED, http://www.ted.com/talks/geoffrey_west_the_surprising_math_of_cities_and_corporations.html (accessed June 10, 2012).

20. Not the change in size as it matures through life, but the average size of the mature animal.

21. West, "Surprising Math of Cities."

Chapter 5. The Energy Substitution Trap

1. T. Jackson, *Prosperity without Growth: Economics for a Finite Planet* (London: Earthscan, 2012).

2. W. S. Jevons, *The Coal Question: An Inquiry Concerning the Progress of the Nation and the Probable Exhaustion of Our Coal-Mines*, 3rd ed., ed. A. W. Flux (1865; repr., London: Macmillan, 1906).

3. Alright, alright, I suppose it's theoretically measurable.

4. H. Groppe, "Peak Oil: Myth vs. Reality," Denver World Oil Conference, November 10–11, 2005, http:// www.energybulletin.net/gpm (accessed February 12, 2007).

5. Jevons, *Coal Question*.

6. C. A. S. Hall and K. A. Klitgaard, *Energy and the Wealth of Nations: Understanding the Biophysical Economy* (New York: Springer, 2012); C. A. S. Hall and J. W. Day Jr., "Revisiting the Limits to Growth after Peak Oil," *American Scientist* 97 (2009): 230–36; N. Gagnon, C. A. S. Hall, and L. Brinker, "A Preliminary Investigation of Energy Return on Energy Investment for Global Oil and Gas Production," *Energies* 2 (2009): 490–503; C. J. Cleveland, R. Costanza, C. A. S. Hall, and R. Kaufmann, "Energy and the United States Economy: A Biophysical Perspective," *Science* 225 (1984): 890–97. C. A. S. Hall, C. J. Cleveland, and R. Kaufmann, *Energy and Resource Quality: The Ecology of the Economic Process* (New York: Wiley, 1986); C. J. Cleveland, "Net Energy from the Extraction of Oil and Gas in the United States," *Energy: The International Journal* 30 (2005): 769–82; M. C. Guilford, P. O'Connor, C. A. S. Hall, and C. J. Cleveland, "A New Long Term Assessment of Energy Return on Investment (EROI) for U.S. Oil and Gas Discovery and Production," *Sustainability* 3 (2011): 1866–87. D. J. Murphy, C. A. S. Hall, and R. Powers, "New Perspectives on the Energy Return on (Energy) Investment (EROI) of Corn Ethanol," *Environment Development and Sustainability* 13 (2010): 179–202.

7. The decrease in electricity production with declining wind speed is dramatic because

the energy available from the wind is not proportional to the square of wind speed, but to the cube.

8. Total US energy consumption is approximately one hundred quads, and total energy collection from photosynthesis by the lower forty-eight states is estimated at eighty quads by David Pimentel, Cornell University. D. Pimentel et al., "Biomass Energy Conversion," *Environmental Biology* 78 (1978): 1–41.

9. Sempra Generation news release, March 18, 2011, http://www.public.sempra.com/newsreleases (accessed July 12, 2012).

10. George Modelski, *World Cities: -3000 to 2000* (Washington, DC: Faros 2000, 2003).

11. National Wind LLC, http://www.nationalwind.com (accessed July 4, 2012); US Energy Information Administration, http://www.eia.gov (accessed July 4, 2012).

12. Well-spotted. This is plagiarized from Mark Twain's "Denial Ain't a River in Egypt."

13. A small number of natural gas plants serve largely to even out production.

14. US Energy Information Administration, http://www.eia.gov (accessed July 4, 2012).

15. Let's face it, Hummers present a safety risk to drivers in smaller (more sensible) cars, and Hummer drivers tend not to be safe drivers. Apparently, they get more than five times the number of traffic tickets than the rest of us. See J. Travers, "Hummer Drivers Lead in Tickets," *Consumer Reports*, October 6, 2009.

16. All miles per gallon (mpg) statements in this section are simplified estimates. The numbers vary depending on the model, there are different values for city versus highway driving, and the estimates vary slightly by country, and so forth. For example, "Hummer" is a broad term. The H1, H2, and H3 Hummers are quite different beasts, so I've estimated somewhere in the middle range. For official figures, please go to the Department of Energy's website, http://www.fueleconomy.gov/ (accessed June 11, 2012).

17. A. Spinella, "Dust to Dust: The Energy Cost of New Vehicles from Concept to Disposal," CNW Marketing Research, Inc., March 2007. I downloaded a PDF of this report from the CNWMR website, http://www.cnwmr.com/nss-folder/automotiveenergy/DUST%20PDF%20VERSION.pdf (accessed June 11, 2012) while researching *The Efficiency Trap*, but it is no longer accessible. For a copy of the report, I can only suggest you contact CNWMR in Brandon, OR. The initial findings of the report were that the Hummer was more efficient than the Prius over the full life cycle of the vehicles. Following lots of criticism, CNWMR rejiggered the numbers, and now the Prius comes out on top. Which is the truth? Well, the Prius is better, especially compared to the monstrous H1 Hummer, but what concerns me here is the fact that it even comes close. Personally, I don't entirely trust any analyses of this type to give a definitive result because the conclusions vary depending on what one considers "the full life cycle," what model is being considered, and what data are available. For an example of the type of criticism leveled against the "Dust to Dust" report, see P. H. Glieck, "Hummer versus Prius: 'Dust to Dust' Report Misleads the Media and Public with Bad Science," Pacific Institute, http://www.evworld.com/library/pacinst_hummerVprius.pdf (accessed January 15, 2012).

18. L. Volsun, "EV Project: Electric Vehicle Journeys Saved a Million Gallons of Gasoline, 8,700 Metric Tons of CO_2 Emissions," *Earth Times*, April 26, 2012, http://www.earthtimes.org/pollution/ev-project-american-electric-vehicles-save-million-gallons-petrol/1952/ (accessed August 20, 2012).

19. D. Gluckman, "2011 Nissan Leaf SL," *Car and Driver*, August 2011, http://www.caranddriver.com/reviews/2011-nissan-leaf-sl-long-term-road-test-review (accessed August 20, 2012).

20. P. F. Drucker, "What Executives Should Remember," *Harvard Business Review* 84 (February 2006).

21. Along with calories from agriculture.

22. J. M. Diamond, *Collapse: How Societies Choose to Fail or Succeed* (New York: Penguin, 2005); T. Homer-Dixon, *The Upside of Down: Catastrophe, Creativity, and the Renewal of Civilization* (Washington, DC: Island Press, 2006); S. G. Hallett and J. Wright, *Life without Oil: Why We Must Shift to a New Energy Future* (Amherst, NY: Prometheus Books, 2011); J. A. Tainter, *The Collapse of Complex Societies* (Cambridge: Cambridge University Press, 1988).

23. Natural gas vehicles are effective and efficient, and further improvements can be made here. I'm discounting them to a certain degree, however, because they really just represent a conversion from one limited source to another and any such conversion would buy very little time in most places (Russia is an exception) and would lose time in others (North America).

24. It is interesting in itself that our fastest growing supply of "gasoline" is ethanol. This is a big worry, since ethanol production is energy neutral, at best, and probably energy negative.

25. D. Pimentel, "Ethanol Fuels: Energy Balance, Economics, and Environmental Impacts Are Negative," *Natural Resources Research* 12 (2003): 127–34; D. Pimentel and T. W. Patzek, "Ethanol Production Using Corn, Switchgrass, and Wood; Biodiesel Production Using Soybean and Sunflower," *Natural Resources Research* 14 (2005): 65–76.

26. R. Hammerschlag, "Ethanol's Energy Return on Investment: A Survey of the Literature 1990–Present," *Environment Science and Technology* 40 (2006): 1744–50.

27. Founder of Mitchell Energy and Development Corporation.

28. There is lots of controversy about the chemicals used in fracking, and for proprietary reasons we don't know exactly what all the various constituents are.

29. And in the next breath they'll tell us that since we have this century-long supply, we can convert our power stations and cars and just about everything else from oil to gas, but that would mean using lots more than we currently do, right? Sorry, but you can't have it both ways. You can't use it more and also expect it to last just as long.

30. They are not discoveries, per se, because we've known they were there for a long time. The discovery is how to exploit them.

31. Hall and Klitgaard, *Energy and the Wealth of Nations*.

32. L. Carroll, *Alice's Adventures in Wonderland* (Wellesley, MA: Branden Books, 1939).

33. Not really "oil," but "keragen," which is a hydrocarbon mix somewhat similar to crude oil—but cruder.

34. Most famously in the movie *Gasland*, directed by Josh Fox (HBO Documentary Films, WOW Publications, 2010).

35. J. Brady, "Face-Off over 'Fracking': Water Battle Brews on Hill," National Public Radio, http://www.npr.org/templates/story/story.php?storyId=104565793&ps=rs (accessed June 26, 2010)

36. P. Spotts, "Fracking Study Sends Alert about Leakage of Potent Greenhouse Gases," *Christian Science Monitor*, February 13, 2012, http://www.csmonitor.com/USA/2012/0213/Fracking-study-sends-alert-about-leakage-of-potent-greenhouse-gas (accessed January 15, 2013).

37. D. McKenzie, "Leaking Gas Mains Help to Warm the Globe," *New Scientist*, September 22, 1990; C. Bylin et al., "New Measurement Data Has Implications for Quantifying Natural Gas Losses from Cast Iron Distribution Mains," *Pipeline and Gas Journal*, September 2009.

38. "The Plowboy Interview with Amory Lovins," *Mother Earth News*, November/December 1977, http://www.motherearthnews.com/Renewable-Energy/1977-11-01/Amory-Lovins.aspx?page=14 (accessed August 23, 2012).

39. This is quite contentious, of course. On the negative (probably impossible) side is the fact that material would need to be stored for tens of thousands of years in an earthquake-free location with no flowing water and various other stable conditions to be completely safe. On the other hand, we can do much better than storing radioactive materials in canisters on site, which is what we do now, so improving on that is a no-brainer, technically, but much harder politically.

40. World Nuclear Association homepage, http://www.world-nuclear.org (accessed August 12, 2012).

41. R. Edwards, "Nuclear Firms Want Special Treatment," *New Scientist*, June 14, 1997.

42. J. Taylor and P. Van Doren, "Nuclear Power in the Dock," *Forbes*, April 5, 2011, http://www.forbes.com/2011/04/04/nuclear-energy-economy-opinions-jerry-taylor-peter-van-doren.html (accessed January 15, 2013).

43. O. Morton, "The Dream That Failed," *Economist*, March 10, 2012.

44. I know: hard to imagine, right?

Chapter 6. The Efficient Food Trap

1. N. Borlaug, Nobel lecture, December 10, 1970, Nobel Institute, Oslo, Norway.

2. This is recognized by various Australian states, whose agriculture departments are known as departments of primary industries.

3. D. A. Pfeiffer, *Eating Fossil Fuels: Oil, Food, and the Coming Crisis in Agriculture* (Gabriola Island, CA: New Society, 2006); P. Roberts, *The End of Food* (New York: Houghton Mifflin, 2008); M. Pollan, *The Omnivore's Dilemma: A Natural History of Four Meals* (New York: Penguin, 2007).

4. *Food, Inc.*, directed by Robert Kenner (River Road Entertainment, Magnolia Pictures, Participant Media, 2008).

5. See the Morrill Land Grant Acts of 1862 and 1870.

6. D. A. Vaccari, "A Phosphorous Famine: The Threat to Our Food Supply," *Scientific American*, June 3, 2009.

7. G. Hardin, "Lifeboat Ethics: The Case against Helping the Poor," *Psychology Today*, September 1974.

8. P. Ehrlich, *The Population Bomb* (New York: Ballantine Books, 1968).

9. Proud to be a Boilermaker. The World Food Prize has been won by Purdue's Phil Nelson (2007) and Gebisa Ejeta (2009).

10. This quote is from an interview with Robert Kenner, the director of *Food, Inc.*, published in the *San Francisco Weekly*, June 12, 2009.

11. Not true as I write. It is August 2012 and the Midwest, especially Indiana, is in yet another one-hundred-year weather event—this time drought.

12. Earl Butz was not without his controversies, and he was known for being outspoken and direct. He eventually resigned after telling an obnoxious racist joke that made it into the press. He was also found guilty of tax evasion, sentenced to jail time, and heavily fined. He lived into his nineties and died a few years ago back here in West Lafayette. (See *Slate*'s commentary on Butz for the nasty, racist joke: Timothy Noah, "Earl Butz, History's Victim: How the Gears of Social Progress Tore Up Nixon's Agriculture Secretary," *Slate*, February 4, 2008, http://www .slate.com/articles/news_and_politics/chatterbox/2008/02/earl_butz_historys_victim.html [accessed January 8, 2013].)

13. R. Paarlberg, "The Culture War over Food and Farming: Who Is Winning?" presentation at Purdue University, April 18, 2012.

14. Co-taught with my wonderful colleagues Steve Weller and Kevin Gibson.

15. "Public Research, Private Gain: Corporate Influence over University Agriculture," Food and Water Watch, April 26, 2012, http://www.foodandwaterwatch.org (accessed September 12, 2012).

16. R. Carson, *Silent Spring* (New York: Houghton Mifflin, 1962).

17. Sorry, Dan. I just can't find a way to get "antidisestablishmentarianism" into the book. I hope this will suffice.

18. The rules are determined for the most part by national agencies. In the United States, independent certifiers are accredited by the US Department of Agriculture (USDA).

19. F. R. Glahe, ed., *Collected Papers of Kenneth Boulding*, vol. 2 (Boulder: Colorado Associated University Press, 1971).

20. Ibid.

21. Ibid.

Part 3: THINKING IN SYSTEMS

Chapter 7. Learning from Nature

1. G. Hardin, *Living within Limits: Economics and Population Taboos* (New York: Oxford University Press, 1995).

2. For some important background reading from some of the leading and original thinkers on the subject of systems thinking, see H. T. Odum and E. C. Odum, *A Prosperous Way Down: Principles and Policies* (Boulder: University of Colorado Press, 2001); C. A. S. Hall and K. A. Klitgaard, *Energy and the Wealth of Nations: Understanding the Biophysical Economy* (New York: Springer, 2012); Donella (Dana) Meadows was the lead author of the Club of Rome's *Limits to Growth* and her unfinished manuscript for *Thinking in Systems* was completed by Diana Wright after her death (D. H. Meadows, *Thinking in Systems: A Primer* [White River Junction, VT: Chelsea Green, 2008]); C. S. (Buzz) Holling has published a veritable library of papers on resilience—here is one of the earlier, fairly broad ones: C. S. Holling, "Understanding the Complexity of Ecological, Economic, and Social Systems," *Ecosystems* 4 (2001): 390–405; and here is one of Holling's more important books: L. H. Gunderson and C. S. Holling, *Panarchy: Understanding Transformations in Human and Natural Systems* (New York: Island Press, 2002).

3. C. R. Darwin, *On the Origin of Species by Means of Natural Selection; or, The Preservation of Favoured Races in the Struggle for Life* (London: John Murray, 1859).

4. For an interesting analysis of this and many other economic oddities, see S. D. Levitt and S. J. Dubner, *Superfreakonomics: Global Cooling, Patriotic Prostitutes, and Why Suicide Bombers Should Buy Life Insurance* (New York: Harper Collins, 2009), the sequel to S. D. Levitt and S. J. Dubner, *Freakonomics: A Rogue Economist Explores the Hidden Side of Everything* (New York: Harper Collins, 2005).

5. Actually, it would be quite difficult to monetize. The size of the porn industry, it turns out, is difficult to assess. Here's one article that addresses it: D. Ackman, "How Big Is Porn?" *Forbes*, May 25, 2001.

6. *Treponema pallidum*—a Spirochete.

7. D. Hayden, *Pox: Genius, Madness, and the Mysteries of Syphilis* (New York: Basic Books, 2003).

8. These last two examples, and many others, are shown in the brilliant, incredibly filmed *Private Life of Plants*, written and presented by the irrepressible David Attenborough (London: BBC TV in association with Turner Broadcasting Systems Inc., 1995).

9. Microbiologists are screaming at me for this simplification. There is much more to prokaryote reproduction than this. Genetic information can be exchanged by plasmids, viruses, and other mechanisms. It's not sexual reproduction in the strictest sense, but it has some of the same effects.

10. J. K. Alexander, *The Mantra of Efficiency: From Waterwheel to Social Control* (Baltimore, MD: Johns Hopkins University Press, 2008).

11. Ponderosa pine (*Pinus ponderosa*), lodgepole pine (*Pinus contorta*), mountain ash (*Eucalyptus regnans*). An interesting note on the mountain ash tree: it is probably a taller tree than the North American redwood, but there are none presently alive at their maximum height thanks to logging.

12. C. Smith, *Media and Apocalypse: News Coverage of the Yellowstone Fires, Exxon Valdez Oil Spill, and Loma Prieta Earthquake* (Westport, CT: Greenwood Press, 1992).

13. We missed a few. Sam was particularly miffed that we didn't see the wolves that were reintroduced in the midnineties, and frankly, the bison, of which there were only a hundred or so at the turn of the twentieth century, were doing a little too well for our liking, causing "bison jams" on the roads.

14. Perhaps they hadn't noticed that ecologists had been doing this kind of analysis for some time and thought they had suddenly come up with something new.

15. . . . solar system . . . Milky Way galaxy . . . universe . . . and so on.

16. The pores in leaves are called stomata.

17. Levitt and Dubner, *Freakonomics*; Levitt and Dubner, *Superfreakonomics*.

18. J. R. Saul, quoted in Meadows, *Thinking in Systems*.

19. The first, simple, and much overused model was known as the Ricker curve or the Ricker stock/recruitment model. W. E. Ricker, "Stock and Recruitment," *Journal of the Fisheries Research Board of Canada* 11 (1954): 559–623.

20. R. Bradbury, "A World without Coral Reefs," *New York Times*, July 13, 2012.

21. Modern-day Uzbekistan and Kazakhstan. The name *Aral Sea* means "Sea of islands." What once were islands became big, dry expanses with three small lakes.

22. M. Wines, "Grand Soviet Scheme for Sharing Water in Central Asia Is Foundering," *New York Times*, December 9, 2002.

23. *Helicoverpa armigera*.

24. Excepting specialty crops grown with supplemental electrical lighting.

25. F. Clements, *Plant Succession: An Analysis of the Development of Vegetation* (Washington, DC: Carnegie Institution of Washington, 1916).

26. *Ammophila arenaria*.

27. Clements, *Plant Succession*.

28. It snuck back over three thousand in August 2012.

Chapter 8. All the Oil in the World

1. D. Floudas et al., "The Paleozoic Origin of Enzymatic Lignin Decomposition Reconstructed from 31 Fungal Genomes," *Science* 336 (2012): 1715–19.

2. They say this all the time. "Without the fungi, bacteria, worms . . . the world would be neck deep in undecomposed trash." Perhaps, in the carboniferous, it was.

3. Or so it would seem. . . . I tried to figure this out, but the data on reserves are problematic; the reserves of natural gas are not known nearly as well as we would need to perform a detailed estimate, and the estimates of recoverable coal reserves range enormously. But I'm getting a figure somewhere around ten trillion tons of CO_2 from now to the end of the fossil fuel interval, with oil and natural gas each accounting for a little less than 30 percent and coal accounting for a little more than 40 percent.

4. The reports of the intergovernmental panel on climate change have a wealth of information, and the summary reports are quite readable. All are available from http://www.ipcc.ch. Some of the better books are: J. Lovelock, *The Revenge of Gaia* (London: Penguin, 2006); T. Flannery, *The Weather Makers: How Man Is Changing the Climate and What It Means for Life on Earth* (New York: Atlantic Monthly Press, 2006); M. Lynas, *Six Degrees: Our Future on a Hotter Planet* (London: Harper Collins, 2007); S. R. Weart, *The Discovery of Global Warming* (Cambridge, MA: Harvard University Press, 2008); J. Hansen, *Storms of My Grandchildren: The Truth about the Coming Climate Catastrophe and Our Last Chance to Save Humanity* (New York: Bloomsbury, 2011).

5. Ben Santer, a specialist in general circulation models, was one of the lead authors for the IPCC reports. He was attacked by various media for distorting data in the IPCC report. No such distortion occurred: Santer is a fine scientist. Some of the mud tends to stick for a while, but it's eventually washed off with some warm water and patience.

6. Carbon dioxide, methane, nitrous oxide, synthetic hydrocarbons such as CFCs and HCFCs. Various other compounds are also greenhouse gases, such as water vapor. The most abundant gases in the atmosphere, nitrogen and oxygen, are not greenhouse gases.

7. K. Emanuel, "Increasing Destructiveness of Tropical Cyclones over the Past 30 Years," *Nature* 436 (2005): 686–88; T. Spencer and I. Douglas, "The Significance of Environmental Change: Diversity, Disturbance, and Tropical Ecosystems," in *Environmental Change and Tropical Geomorphology*, ed. I. Douglas and T. Spencer (London: Allen & Unwin, 1985).

8. Well worth looking into though. Here is a 2009 Royal Society (UK) report to get you started: *Geoengineering the Climate: Science, Governance, and Uncertainty*, policy document 10/09, 2009, http://www.royalsociety.org/uploadedFiles/Royal_Society_Content/policy/publications/2009/8693.pdf (accessed September 2, 2012).

9. G. Hardin, "The Tragedy of the Commons," *Science* 162 (1968): 1243–48.

10. We were initially warned that we should try to keep carbon dioxide concentrations below 350 ppm. We passed that threshold a while back. We were then warned that a carbon dioxide concentration of 400 ppm was the "point of no return," which is a bit of a problem because this is probably a reasonable number and we've just about reached that now. We tend to focus on what kind of effects we are likely to see by century's end, and so when people speak about catastrophic 4°C increases in temperature, they are still hanging on to the hope that we can put a lid on this thing. I really don't think we can. The worst scenarios are inevitable when we look centuries forward rather than just decades. In geological terms there seems to be no

obvious reason why we shouldn't expect carbon dioxide concentrations in the thousands of parts per million and global temperatures last seen fifty million years ago when palm trees grew in Siberia, crocs roamed Antarctica, and no large mammals could survive on the parts of the planet where most of us live today.

11. W. W. Ruddiman, *Plows, Plagues, and Petroleum: How Humans Took Control of Climate* (Princeton, NJ: Princeton University Press, 2005).

12. Water vapor is actually the greenhouse gas that is found in the atmosphere in the largest concentrations, and its effects cannot be ignored, though they are very hard to predict. Water vapor is a rather weak greenhouse gas, but since there is so much of it, it is a major contributor to the greenhouse effect. Global warming will increase rates of evaporation, sending more water into the atmosphere, but this will cause more cloud formation, bringing the water back down. Understanding cloud effects is tough. They reflect incoming solar radiation, cooling the planet, but they also trap outgoing heat, warming it. The balance of cooling versus warming depends on the height and density of the cloud. Water is neither made nor destroyed in the environment (actually not true—plants break up water, but they put it back together again). Rather, water simply cycles at different speeds and spends time in different pools (ice caps, oceans, lakes, aquifers, vapor, and so forth). Global warming will certainly mess with the water cycle, but what the net effects of this will be are very hard to predict. And there's nothing much we can do to manage atmospheric water vapor directly.

13. G. Hardin, *Filters against Folly: How to Survive Despite Economists, Ecologists, and the Merely Eloquent* (New York: Viking, 1985); S. G. Hallett and J. Wright, *Life without Oil: Why We Must Shift to a New Energy Future* (Amherst, NY: Prometheus Books, 2011).

14. CFC: chlorofluorocarbon; HCFC: hydrochlorofluorocarbon; HFC: hydrofluorocarbon. These are a suite of somewhat similar molecules all used in heat exchange and air conditioning. They have similar chemical characteristics and they are all, to a greater or lesser degree, greenhouse gases (the CFCs are the most potent). Their half-life in the atmosphere is rather long in most cases. HCFC-123 has the shortest half-life, and the CFCs have extremely long half-lives. The big difference is that while CFCs chomp up stratospheric ozone, the HCFCs and HFCs do not, or at least they do so to a much lesser degree. An example trade name is DuPont's Freon, used for various different compounds.

15. National Oceanic and Atmospheric Administration, National Aeronautics and Space Administration, United Nations Environment Programme, World Meteorological Organization, European Commission, "Scientific Assessment of Ozone Depletion: 2006," Executive Summary, reprinted from *World Meteorological Organization Global Ozone Research and Monitoring Project—Report No. 50*, http://www.esrl.noaa.gov/csd/assessments/ozone/2006/chapters/executivesummary.pdf (accessed January 14, 2013). The ozone hole is on the mend, which is a really good thing, keeping that pesky, mutagenic UV under control. It should be pretty much fixed not long after midcentury.

16. E. Rosenthal and A. W. Lehren, "Relief in Every Window, but Global Worry Too," *New York Times*, June 20, 2012.

17. G. J. M. Velders et al., "The Large Contribution of Projected HFC Emissions to Future Climate Forcing," *Proceedings of the National Academy of Science* 106 (2009): 10949–54.

18. P. Spotts, "Fracking Study Sends Alert about Leakage of Potent Greenhouse Gases," *Christian Science Monitor*, February 13, 2012, http://www.csmonitor.com/USA/2012/0213/Fracking-study-sends-alert-about-leakage-of-potent-greenhouse-gas (accessed January 15, 2013).

19. The first rice farmers also began to change the climate around the same time. The cultivation of paddy rice creates anoxic conditions in which methane-releasing microbes thrive. Ruddiman, *Plows, Plagues, and Petroleum.*

20. National Agricultural Statistics Service, http://www.nass.usda.gov/ (accessed July 2, 2012).

21. Another irony is that the rich, corn-based feed makes the cattle sick, and we need to treat them for that, too.

Chapter 9. Efficient Business and Economy Traps

1. R. Levett, "Rebound and Rational Public Policy Making," in *Energy Efficiency and Sustainable Consumption: The Rebound Effect*, ed. H. Herring and S. Sorrell (Basingstoke, UK: Palgrave Macmillan, 2009).

2. Which is why *Life without Oil* and *The Efficiency Trap* are so impeccably error free— well, that and the thousands of corrections made by the editors at Prometheus Books . . . and I'm hoping you missed the few that inevitably made it through.

3. "No End to Paperwork," *Global Trends*, September 18, 2000, available at World Resources Institute, http://www.wri.org/publication/content/8440 (accessed January 14, 2013).

4. M. Martin, *Business Efficiency for Dummies* (New York: John Wiley, 2013); N. J. Sayer and B. Williams, *Lean for Dummies* (New York: John Wiley, 2012); N. Grunden, *The Pittsburgh Way to Efficient Healthcare: Improving Patient Care Using Toyota-Based Methods* (New York: CRC Press, 2007); A. K. Robertson and W. Proctor, *Work a Four-Hour Day: Achieving Business Efficiency on Your Own Terms* (New York: Harper Collins, 1996); M. M. Shirley, *Thirsting for Efficiency: The Economics and Politics of Urban Water System Reform* (Maryland Heights, MO: Elsevier, 2002).

5. OK, so there might be some sour grapes here, since so many of them are also selling more copies than me.

6. T. Freidman, *The World Is Flat: A Brief History of the Twenty-First Century* (New York: Farrar, Strauss and Giroux, 2005).

7. J. P. Womack, D. T. Jones, and D. Roos, *The Machine That Changed the World: Based on the MIT Five-Million-Dollar Five-Year Study on the Future of the Automobile* (New York: Simon & Schuster, 1990); J. P. Womack and D. T. Jones, *Lean Thinking: Banish Waste and Create Wealth in Your Corporation* (New York: Simon & Schuster, 1996).

8. J. K. Liker, *The Toyota Way: Fourteen Management Principles from the World's Greatest Manufacturer* (New York: McGraw Hill Professional, 2004).

9. M. L. George, *Lean Six Sigma: Combining Six Sigma Quality with Lean Speed* (New York: McGraw Hill Professional, 2002).

10. Not actually the biggest employer in America (that's the Department of Defense), but the biggest corporate employer, and actually the number three overall employer in the world after the US military and the Chinese military. McDonalds is not too far behind. Ruth Alexander, "Which Is the World's Biggest Employer?" BBC News, http://www.bbc.co.uk/news/magazine-17429786 (accessed June 8, 2012) and the *Economist*, June 2012.

11. C. Fishman, *The Walmart Effect: How the World's Most Powerful Company Really Works—and How Its Transforming the American Economy* (New York: Penguin, 2006).

12. G. Hardin, *Filters against Folly: How to Survive Despite Economists, Ecologists, and the Merely Eloquent* (New York: Viking, 1985).

13. R. U. Ayres and B. Warr, "Energy Efficiency and Economic Growth: The Rebound Effect as a Driver," in Herring and Sorrell, *Energy Efficiency and Sustainable Consumption*.

14. BTU = British Thermal Units = approximately 1 kJ. K. Ehrhardt-Martinez and J. A. Laitner, *The Size of the US Energy Efficiency Market: Generating a More Complete Picture*, Report No. E083 (Washington, DC: American Council for an Energy-Efficient Economy, 2008). Available online at http://www.aceee.org/publications (accessed August 23, 2011).

15. S. Kuznets, "Economic Growth and Income Inequality," *American Economic Review* 45 (1955): 1–28.

16. M. Friedman, *There Is No Such Thing as a Free Lunch* (Chicago, IL: Open Court, 1975).

17. R. K. Kaufmann, "The Mechanisms for Autonomous Increases in Energy Efficiency: A Cointegration Analysis of the US energy/GDP ratio," *Energy Journal* 25 (2004): 63–86.

18. A. Huxley, *Tomorrow and Tomorrow and Tomorrow and Other Essays* (New York: Harper Brothers, 1956).

19. You must check out Eddie Izzard on this. "We Stole Countries . . . with the Cunning Use of Flags," YouTube, http://www.youtube.com/watch?v=uEx5G-GOS1k (accessed January 14, 2013).

20. Mitch Daniels, quoted in Tom Cook, "Indiana Gov. Mitch Daniels' Legacy: Privatization," *Indianapolis Star*, December 29, 2012, available at Indystar.com, http://www.indystar.com/article/20121229/NEWS05/212290326/Indiana-Gov-Mitch-Daniels-legacy-Privatization (accessed January 24, 2013).

21. It was an odd committee, and I think I disrupted proceedings a fair bit with all my questions, which is a shame because I very much liked all the people on the committee.

22. "Preservation of Antibiotics for Medical Treatment Act," Union of Concerned Scientists, last revised January 4, 2013, http://www.ucsusa.org/food_and_agriculture/solutions/wise_antibiotics/pamta.html (accessed January 14, 2013).

Chapter 10. The Ecology of Collapse

1. I. Asimov, *Foundation* (New York: Doubleday, 1951). *Foundation* was apparently inspired by Edward Gibbon's *Decline and Fall of the Roman Empire*.

2. R. Kurzweil, *The Age of Spiritual Machines: When Computers Exceed Human Intelligence* (New York: Penguin, 2000).

3. Modern technology, that is: ignoring the wheel, fire, agriculture, and so forth.

4. G. E. Moore, "Cramming More Components onto Integrated Circuits," *Electronics*, April 19, 1965.

5. R. Kurzweil, *The Singularity Is Near: When Humans Transcend Biology* (New York: Penguin, 2006).

6. My version is based on Sissa ben Dahir and King Shihram, in India circa the thirteenth century, but there are various versions, one with rice rather than wheat, another from ancient Rome.

7. R. Dawkins, *The Extended Phenotype: The Long Reach of the Gene* (Oxford: Oxford University Press, 1999).

8. K. Kelly, *What Technology Wants* (New York: Penguin, 2010).

9. Resources that can be accessed by more than one individual and are not infinite or infinitely replaceable and can, therefore, be depleted.

10. G. Hardin, "The Tragedy of the Commons," *Science* 162 (1968): 1243–48. G. Hardin, *Filters against Folly: How to Survive Despite Economists, Ecologists, and the Merely Eloquent* (New York: Viking, 1985).

11. Nobel Prize in Economic Sciences, 2009. Professor Ostrom died of pancreatic cancer in June 2012. Her seminar at Purdue was given on April 16, 2012.

12. E. Ostrom, *Governing the Commons: The Evolution of Institutions for Collective Action* (Cambridge: Cambridge University Press, 1993); E. Ostrom et al., *Rules, Games, and Common-Pool Resources* (Ann Arbor, MI: University of Michigan Press, 1994); B. Vollan and E. Ostrom, "Cooperation and the Commons," *Science* 330 (2012): 923–24.

13. Why is it called a fishery? Shouldn't it be a crustaceary?

14. E. F. Schumacher, "Small Is Beautiful," an essay in *Radical Humorist* 37 (April 1973): 22.

15. Right, OK, I shouldn't speak so glibly of speakers with their little personal "save the world" scenarios. Methinks there might be the occasional listener out there who sees me that way.

16. In fact, Tim Jackson (in *Prosperity without Growth*) estimates that if we were to support the theoretical nine billion population at a decent level of wealth, we would need fifteen earths.

17. Garrett Hardin, in a 1985 talk titled "An Ecolate View of the Human Predicament," which was eventually developed into *Filters against Folly*. See the Garrett Hardin Society, http://www.garretthardinsociety.org/articles/art_ecolate_view_human_predicament.html (accessed August 14, 2012).

18. D. R. Klein, "The Introduction, Increase, and Crash of Reindeer on St. Matthew Island," *Journal of Wildlife Management* 32 (1968): 350–67.

19. P. Hawken, H. Lovins, and A. Lovins, *Natural Capitalism: The Next Industrial Revolution* (New York: Hatchette Book Group, 2010).

20. The Dine are often known as the Navajo, but they call themselves Dine. The Hisatsinom were called the Anasazi (by the Dine) but are called the Hisatsinom by the Hopi and other tribes that are among their descendants.

21. More complete quote: "One might argue . . . that at every stage of history our concern must be to dismantle those forms of authority and oppression that survive from an era when they might have been justified in terms of the need for security or survival or economic development but that now contribute to—rather than alleviate—material and cultural deficit. If so, there will be no doctrine of social change fixed for the present and future, nor even, necessarily, a specific and unchanging concept of the goals towards which social change should tend. Surely our understanding of the nature of man or of the range of viable social forms is so rudimentary that any far-reaching doctrine must be treated with great skepticism, just as skepticism is in order when we hear that 'human nature' or 'the demands of efficiency' or 'the complexity of modern life' requires this or that form of oppression or autocratic rule." Noam Chomsky, "Notes on Anarchism," chapter 7 in *The Essential Chomsky* (New York: New Press, 2008).

22. J. M. Diamond, *Collapse: How Societies Choose to Fail or Succeed* (New York: Penguin, 2005).

23. J. Van Tilburg, *Easter Island: Archaeology, Ecology, and Culture* (Washington, DC: Smithsonian Institution Press, 1994).

24. Military personnel statistics, United States Department of Defense, http://siadapp.dmdc.osd.mil/personnel/MILITARY/miltop.htm (accessed January 15, 2013).

25. So is it "Lean Finely Textured Beef" or "Ammonia-Treated Burger Extender"? You be the judge. M. Bittman, "The Pink Menace," *New York Times*, April 3, 2012.

26. Correctional populations in the United States, 2011, Bureau of Justice Statistics, http://bjs.ojp.usdoj.gov/index.cfm?ty=tp&tid=11 (accessed April 12, 2012).

27. G. J. Bryjak, *Adirondack Daily Enterprise*, April 15, 2000.

28. N. Klein, *The Shock Doctrine: The Rise of Disaster Capitalism* (New York: Macmillan, 2008).

29. T. Frank, *What's the Matter with Kansas? How Conservatives Won the Heart of America* (New York: Macmillan, 2005).

30. Steven Biko, speech at the Cape Town Conference on Inter-Racial Studies, Abe Bailey Institute for Inter-racial Studies, Cape Town, South Africa, January 1971.

31. M. A. Johnson, "The Culture of Einstein," MSNBC, 2005, available online at http://www.msnbc.msn.com/id/7406337/ (accessed July 9, 2012).

32. Theoretical sheep, that is. We don't have hundreds of sheep in the classroom. . . . This is not New Zealand.

33. There are often one or two students who refuse to participate in the tragedy, seeing the need to preserve the pasture from the get-go, but they are the first to starve and their removal from the game focuses the minds of the rest.

Part 4: BEYOND EFFICIENCY

Chapter 11. Resilience: Beyond Sustainability

1. G. Hardin, *Living within Limits: Ecology, Economics, and Population Taboos* (Oxford: Oxford University Press, 1993).

2. They didn't work, of course.

3. OK, so this is not strictly true. There are certain systems powered by geothermal energy, such as hot springs and underwater thermal vents.

4. The relative abundance of different building blocks has changed. The early atmosphere was rich in good carbon mixtures and devoid of oxygen.

5. Hurricanes, for example, are fairly frequent in some areas but not in others.

6. T. Jackson, *Prosperity without Growth: Economics for a Finite Planet* (London: Earthscan, 2012).

7. Especially the earliest, such as Aristotle, but also early-modern thinkers such as John Locke and even, to a degree, Adam Smith.

8. Such as Adam Smith, Jean-Baptiste Say, David Ricardo, Karl Marx.

9. Reinterpretation of Adam Smith and others by Milton Friedman, John Keynes, and others.

10. H. E. Daly, *Steady State Economics: Second Edition with New Essays* (New York: Island Press, 1991).

11. E. F. Schumacher, *Small Is Beautiful: Economics as if People Mattered* (New York: Harper Collins, 2010).

12. Jackson, *Prosperity without Growth*.

13. C. A. S. Hall and K. A. Klitgaard, *Energy and the Wealth of Nations: Understanding the Biophysical Economy* (New York: Springer, 2012).

14. G. O. Barney, *Global 2000 Report to the President*, vol. 3 (Washington, DC: Government Printing Office, 1980).

15. This is not a true feedback because the depletion of the resource does not trigger the response.

16. Tim Flannery points out that the great agricultural regions of the world are located in areas with newly formed soils, primarily from glaciation events. If you want to know how productive a piece of land is, you need only ask its inhabitants, "How good was your last glaciation?" T. Flannery, *The Weather Makers: How Man Is Changing the Climate and What It Means for Life on Earth* (New York: Atlantic Monthly Press, 2006).

17. Sudden oak death has had a huge effect in forests, particularly on the West Coast. It is caused by an oomycete pathogen, *Phytophthora ramorum*. For more information, contact "laughing" Gail Ruhl in Purdue's disease diagnostics lab at http://www.ppdl.purdue.edu/ppdl/SOD.html.

18. Some Gini coefficients: Japan 24.9, Sweden 25.0, Germany 28.3, France 32.7, Pakistan 33.0, Canada 33.1, United Kingdom 36.0, Iran 43.0, United States 46.6, Mexico 54.6, South Africa 57.8, Namibia 70.7. Source: Hall and Klitgaard, *Energy and the Wealth of Nations*.

19. I must confess, I don't know who the heck Gringo Stars is. He/she is cited in Derrick Jensen's brilliant first volume of *Endgame*. D. Jensen, *The Problem of Civilization: Volume 1 of Endgame* (New York: Seven Stories Press, 2006).

Chapter 12. Strengthening Vital Systems

1. Stephen Colbert, White House Correspondents' Association Dinner, Hilton Washington hotel, Washington, DC, April 29, 2006.

2. "Municipal Solid Waste," US Environmental Protection Agency, http://www.epa.gov/epawaste/nonhaz/municipal/index.htm (accessed August 16, 2012).

3. J. H. Kunstler, *The Long Emergency: Surviving the End of Oil, Climate Changes, and Other Converging Catastrophes of the Twenty-First Century* (New York: Grove Press, 2005). See also J. H. Kunstler, *The Geography of Nowhere: The Rise and Decline of America's Man-Made Landscape* (New York: Simon & Schuster, 1994), and *The End of Suburbia: Oil Depletion and the Collapse of the American Dream*, directed by Gregory Greene (Electric Wallpaper Company, Ontario, Canada, 2004).

4. But I could have done without the psychic powers thing.

5. W. McDonough and M. Braungart, *Cradle to Cradle: Remaking the Way We Make Things* (New York: North Point Press, 2002).

6. Ibid.

7. H. D. Thoreau, "Essay: Life without Principle," *Atlantic Monthly*, October 1963.

8. American anthropologist active in the 1960s and 1970s. This quote has been used a number of times in recent sociological literature, attributed to her, but I can't find the original source.

9. Paul Brown, "Woking Shines in Providing Renewable Energy," *Guardian*, http://www.guardian.co.uk/environment/2004/jan/26/energy.renewableenergy (accessed July 23, 2012).

10. "Victorian Community Goes It Alone on Wind Farm," *ABC News*, July 27, 2008, http://www.abc.net.au/news/2008-07-25/victorian-community-goes-it-alone-on-wind-farm/ 452608. And the project has its own web site: http://hepburnwind.com.au (accessed July 23, 2012).

11. J. Thomas, "Danish Island Is Energy Self-Sufficient," *MetaEfficient*, http://www.metaefficient.com/renewable-power/danish-island-is-energy-self-sufficient.html (accessed January 14, 2013).

12. Transition Network, http://www.transitionnetwork.org/.

13. R. Hopkins, *The Transition Handbook: From Oil Dependency to Local Resilience* (White River Junction, VT: Chelsea Green, 2008); S. Chamberlin, *The Transition Timeline for a Local, Resilient Future* (White River Junction, VT: Chelsea Green, 2009).

14. B. Mollison and R. M. Slay, *Introduction to Permaculture* (Stanley, Australia: Tagari, 1994); B. Mollison, *Permaculture: A Designer's Manual* (Stanley, Australia: Tagari, 1988); D. Holmgren, *Permaculture: Principles and Pathways beyond Sustainability* (Hepburn, Australia: Holmgren Design Services, 2002). David's webpage is well worth a visit: http://www.holmgren .com.au/ (accessed September 2, 2012).

15. Energy Descent Plans for Kinsale, Ireland: R. Hopkins, ed., "Kinsale 2021: An Energy Action Plan," Kinsale Further Education College, 2005, http://www.transitionculture.org/wp -content/uploads/KinsaleEnergyDescentActionPlan.pdf (accessed August 14, 2012); for Totnes, UK: Transition Town Totnes, "Transition in Action, Totnes 2030, an Energy Descent Action Plan," totnesedap.org.uk/; for Bloomington, IN: Bloomington Peak Oil Task Force, "Redefining Prosperity: Energy Descent and Community Resilience," http://www.bloomington.in.gov/ media/media/application/pdf/6239.pdf (accessed August 14, 2012); for San Francisco, CA: San Francisco Peak Oil Preparedness Task Force, *Report of the San Francisco Peak Oil Preparedness Task Force*, http://www.postcarboncities.net/files/PeakOilTaskForceReport031709.pdf (accessed August 14, 2012).

16. P. Hampel and V. Schremp Hahn, "Ferguson Resident Wins Fight for Front Yard Vegetable Garden," *St. Louis Post Dispatch*, July 27, 2012, http://www.stltoday.com/news/ local/metro/ferguson-resident-wins-fight-for-front-yard-vegetable-garden/article_a5452134 -2a26-5d c5-9c16-0c1c212c8ff3.html (accessed August 23, 2012).

17. Eddie Izzard, "First Thanksgiving," YouTube, http://www.youtube.com/watch?v= qAOQtp-3b48 (accessed January 14, 2013).

18. F. D. Roosevelt, "Letter to All State Governors on a Uniform Soil Conservation Law," American Presidency Project, http://www.presidency.ucsb.edu/ws/?pid=15373 (accessed January 23, 2013).

19. Clay Bottom Farm, http://www.claybottomfarm.com (accessed August 20, 2012).

20. CSA is short for community-supported agriculture. The basic idea is to sell baskets of food in advance to help generate capital to start the season. The customer pays, say $600 at the beginning of the growing season and then receives a $25 basket per week from May through October.

21. Or it could be the Contrarian Farmer Society, inspired by Wendell Berry's brilliant poem (a personal favorite) "The Contrariness of the Mad Farmer," available in W. Berry, *Farming: A Hand Book* (Berkeley, CA: Counterpoint Press, 2011).

22. Oddly enough, Ron Paul is a big supporter of raw milk.

23. Growing Power, http://www.growingpower.org (accessed August 30, 2012); P-Patch, http://www.seattle.gov/neighborhoods/ppatch/ (accessed August 30, 2012).

24. A quick shout out to the student farm club, Full Circle Agriculture at Purdue, http:// www.ag.purdue.edu/programs/student farm/ (accessed January 14, 2013).

25. Ben Hartman, e-mail to author, August 23, 2012.

Chapter 13. Building Resilient Communities

1. V. Havel, speech accepting membership in the Academy of Ethics and Political Science at the Institut de France, November 13, 1992.

2. Garrett Hardin, "Garrett Hardin Quotes," Garrett Hardin Society, http://www.garrett hardinsociety.org/info/quotes.html (accessed January 23, 2013).

3. M. Gladwell, *The Tipping Point: How Little Things Can Make a Big Difference* (New York: Back Bay Books, 2000), p. 7. See also S. D. Levitt and S. J. Dubner, *Superfreakonomics: Global Cooling, Patriotic Prostitutes, and Why Suicide Bombers Should Buy Life Insurance* (New York: Harper Collins, 2009), which is the sequel to S. D. Levitt and S. J. Dubner, *Freakonomics: A Rogue Economist Explores the Hidden Side of Everything* (New York: Harper Collins, 2005); and M. Gladwell, *Outliers: The Story of Success* (New York: Hatchette, 2008).

4. C. Heath and D. Heath, *Switch: How to Change Things When Change Is Hard* (Toronto: Random House, 2010).

5. Winston S. Churchill, speech to the House of Commons, November 11, 1947.

6. Ruth Bader Ginsberg, Dissenting in *Bush v. Gore*, 2000. The full text of her opinion can be found at Legal Information Institute, Cornell University Law School, http://www.law .cornell.edu/supct/html/00-949.ZD2.html (accessed January 23, 2013).

7. G. Orwell, *Nineteen Eighty-Four* (London: Secker and Warburg, 1949).

8. H. E. Daly, "Economics in a Full World," *Scientific American* 293 (2005).

9. C. McCarthy, *The Road* (New York: Random House, 2010).

10. A. Huxley, *Tomorrow and Tomorrow and Tomorrow and Other Essays* (New York: Harper and Brothers, 1956).

Epilogue: The Key That Unlocks the Efficiency Trap

1. OK, so not entirely true, obviously. Continents are eventually buried, a hot spring might dry up, and so forth.

2. Which we then logged, plowed, and drained over the last two hundred years.

INDEX

Italicized page numbers indicate images.